WORKED EXAMPLES

APPLIED MATHS

by

C. E. Walsh

CELTIC REVISION AIDS

CELTIC REVISION AIDS
Lincoln Way, Windmill Road,
Sunbury on Thames, Middlesex

First published in this edition 1979

ISBN 017 751157 5

Printed in Hong Kong

ACKNOWLEDGEMENTS

The publishers wish to thank the following for permission to use actual examination questions in this book:

The Associated Examination Board

The Delegates of Local Examinations, University of Oxford.

1. To find the moment of inertia of a uniform solid right circular cylinder about a diameter of one end.

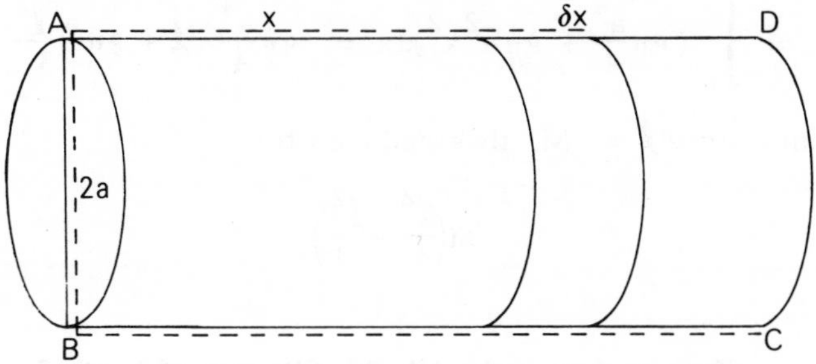

In the diagram, ABCD is the cylinder, AB being a diameter of one end. Let the mass of the cylinder be M, its length ℓ, and radius of cross-section a. Let ρ be the density of the cylinder. Divide the cylinder into thin circular discs and consider one of length δx, at distance x from AB. The mass, δm, of this disc will be given by $\delta m = \rho \pi a^2 . \delta x$.

The moment of inertia of this disc about any of its diameters will therefore be:

$$\delta m . \frac{a^2}{4}$$

by the usual formula:

$$= \rho \pi \frac{a^4}{4} \delta x$$

Now AB is distant x from the centroid of this disc. Consequently, by the theorem of parallel axes, the moment of inertia of the disc about AB:

$$= \rho \pi \frac{a^4}{4} . \delta x + \delta m . x^2$$

$$= \left(\rho\pi\frac{a^4}{4} + \rho\pi a^2x^2\right)\delta x$$

Now summing for all the discs, we see that the required moment of inertia of the cylinder about AB:

$$= \int_0^{\ell}\left(\rho\pi\frac{a^4}{4} + \rho\pi a^2x^2\right)dx = \rho\pi\frac{a^4}{4}.\ell + \rho\pi\frac{a^2\ell^3}{3}$$

Since $\rho\pi a^2\ell = M$, this reduces to:

$$M\left(\frac{a^2}{4} + \frac{\ell^2}{3}\right)$$

2. **Three uniform rods, AB, BC, CD, each of length 2a and weight W, are smoothly jointed together at B and C. The end A is smoothly hinged to a fixed point on a rough horizontal bar and the end D is fastened to a small light ring which is threaded on the bar, the whole system lying in a vertical plane. The system is in equilibrium with BC horizontal and both AB and CD making an angle θ with the vertical. Find the force of friction. at D.**

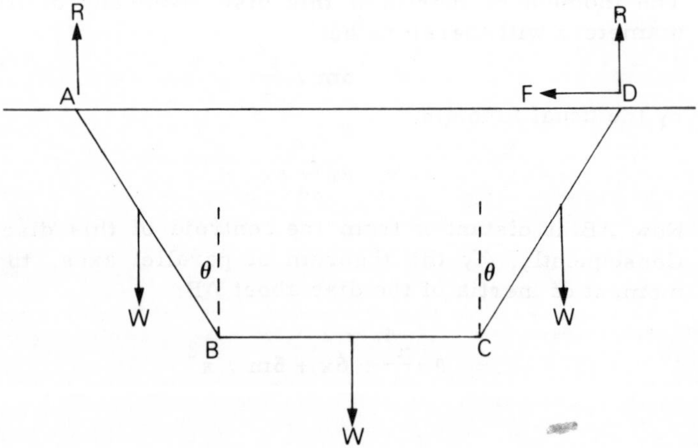

Given that the coefficient of friction at D is 2/3, prove that equilibrium is possible only if $\theta \leqslant \frac{1}{4}\pi$.

(Ox)

Let R and F be the normal or vertical reaction and force of friction respectively, acting on the rod CD at D. From symmetry the normal reaction at A will also be R. Resolving vertically for the equilibrium of the system formed by the three rods, we get:

$$2R = 3W, \text{ so } R = \frac{3W}{2} \qquad (1)$$

To find F, take moments about C for the equilibrium of the rod CD.

$$Wa \sin\theta = R \cdot 2a \sin\theta + F \cdot 2a\cos\theta$$

$$= 3Wa \sin\theta + F \cdot 2a \cos\theta \text{ from } (1)$$

Thus: $$F = -W\tan\theta \qquad (2)$$

the minus sign indicating that F really acts towards the right and so force of friction at D is W tan θ along AD. If μ is the coefficient of friction, equilibrium is possible only if:

$$\frac{F}{R} \leqslant \mu$$

Here $\mu = \frac{2}{3}$. Hence, substituting for F and R:

$$\frac{W\tan\theta}{3\frac{W}{2}} \leqslant \frac{2}{3} \text{ for equilibrium,}$$

i.e.

$$\tan\theta \leqslant 1, \text{ or } \theta \leqslant \frac{\pi}{4}$$

3. **The figure shows a uniform rod AB, of weight W and length 2ℓ, in limiting equilibrium with its end B on a rough horizontal plane. An inextensible light string is attached at the end A, passes over a smooth peg P and carries a weight W at its free end. The peg is vertically above B and at the same level as the mid-point of the rod; the rod and the peg lie in the same vertical plane. Prove that AP is inclined to the horizontal at an angle of 30° and calculate:**

(a) the angle ABP
(b) the value of the coefficient of friction

(AEB)

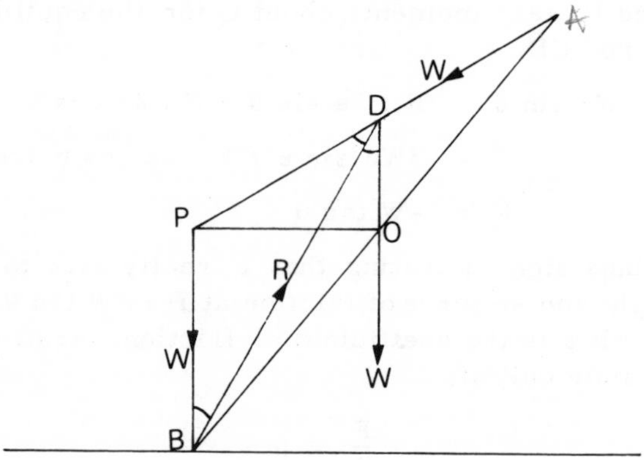

Let O be the mid-point of the rod, then PO will be horizontal and the weight W of the rod will act through O. Let its line of action meet AP in D. Let the total reaction of the ground on the rod at B be R. Since the peg is smooth, the tension in the portion of string AP will also be W. There are three forces acting on the rod and keeping it in equilibrium, two equal forces W whose lines of action meet at D, and R. Therefore the line of action of R must also pass through D, i.e. R acts along BD. Moreover, BD must bisect the angle PDO between these two equal forces W, therefore $\angle PDB = \angle BDO$. But, as equilibrium is limiting, $\angle PBD = \lambda$, the angle of friction. Hence also $\angle BDO = \angle PBD$ (alternate angles) $= \lambda$. Thus $\angle PDB = \angle PBD = \lambda$, so that $PB = PD$. Now since O is the middle point of AB, and OD is parallel to PB, from geometry $DO = \frac{1}{2}PB = \frac{1}{2}PD$. Consequently, as $\angle POD = 90^{\circ}$:

$$\sin APO = \frac{OD}{PD} = \frac{1}{2}, \therefore \angle APO = 30^{\circ}$$

Furthermore $\angle PDO = 2\lambda$, i.e. $2\lambda = 60^{\circ}$, $\lambda = 30^{\circ}$. The coefficient of friction:

$$\mu = \tan\lambda = \frac{1}{\sqrt{3}} = \frac{\sqrt{3}}{3}$$

Finally, from triangle OPB:

$$\tan ABP = \frac{OP}{PB} = \frac{OP}{PD} = \cos 30^{\circ} = \frac{\sqrt{3}}{2}$$

so: $$\angle ABP = \tan^{-1}\left(\frac{\sqrt{3}}{2}\right)$$

4. A uniform cube, of edge 2a and weight W_1, and a uniform hemisphere, of radius a and weight W_2, rest on a rough horizontal table. The hemisphere touches one of the vertical faces of the cube at the central point of that face. This point of contact, the centres of mass of the cube and of the hemisphere are in one vertical plane, which is that of the section shown in the figure. The plane face of the hemisphere is inclined to the horizontal at an angle α. The coefficients of friction between the cube and the table and between the hemisphere and the table are μ_1 and μ_2 respectively. There is no friction between the cube and the hemisphere. Prove that equilibrium is possible only if 3/8 sin α is less than or equal to the least of:

(Ox)

$$\frac{\mu_1 W_1}{W_2}, \mu_2, \text{ and } \frac{W_1}{W_2}$$

(Ox)

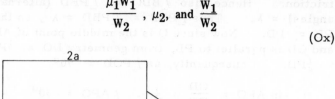

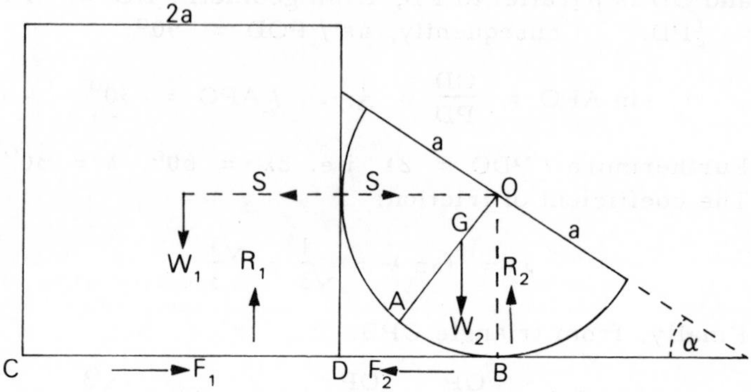

Let O be the centre of the plane face of the hemisphere, OA the radius bisecting the hemisphere. Then G the centre of gravity of the hemisphere is a point G along OA so that OG = 3/8a. Let CD be the line in which this section of the cube meets the table, B the point where hemisphere touches the table. Denote by R_1 and R_2 the normal reactions of the table on cube and sphere respectively, and by F_1 and F_2 the respective forces of friction. Since there is no friction between the cube and the hemisphere, the reaction between them will be normal to both, and thus act horizontally; denote it by S.

Resolving horizontally for the equilibrium of the whole system:

$$F_1 = F_2$$

Resolving horizontally for the equilibrium of the cube:

$$S = F_1$$

$$\therefore \quad F_1 = F_2 = S \qquad (1)$$

Resolving vertically first for the equilibrium of the cube, then for that of the sphere:

$$R_1 = W_1 \qquad (2)$$

$$R_2 = W_2 \qquad (3)$$

Taking moments round B for the equilibrium of the hemisphere:

$$S \cdot a = W_2 \cdot OG \sin \alpha = W_2 \cdot \frac{3}{8} a \sin \alpha, \text{ so:}$$

$$S = \frac{3}{8} \cdot W_2 \sin \alpha \qquad (4)$$

For equilibrium there must be no sliding of cube or hemisphere and so:

$$\frac{F_1}{R_1} \leqslant \mu_1 \quad ; \quad \frac{F_2}{R_2} \leqslant \mu_2$$

must be fulfilled, i.e. from (1) (2) and (3):

$$\frac{S}{W_1} \leqslant \mu_1 \quad ; \quad \frac{S}{W_2} \leqslant \mu_2$$

Thus substituting for S from (4):

$$\frac{3}{8} \sin \alpha \leqslant \frac{\mu_1 W_1}{W_2} \quad ; \quad \frac{3}{8} \sin \alpha \leqslant \mu_2$$

In addition, taking moments round C for the equilibrium of the cube:

$$Sa + \text{moment of } R_1 \text{ round C} = W_1 a, \text{ so that:}$$

$$Sa \leqslant W_1 a \quad ; \quad \text{or: } S \leqslant W_1$$

$$\therefore \quad \frac{3}{8} W_2 \sin \alpha \leqslant W_1 \quad ; \quad \frac{3}{8} \sin \alpha \leqslant \frac{W_1}{W_2}$$

Hence $\frac{3}{8}\sin a$ is less than or equal to the least of:

$$\frac{\mu_1 W_1}{W_2}, \ \mu_2, \text{ and } \frac{W_1}{W_2}$$

5. **The end A of a light inelastic string AB of length 2a is fixed and the end B is attached to a uniform rod BC, of length 2a and weight W. The end C of the rod rests on a fixed rough horizontal table below A, the coefficient of friction between the rod and the table being μ. If the rod is in equilibrium with C vertically below A, and if angle BAC is then α, find the tension in the string in terms of W and α, and prove that μ cannot be less than:**

$$\frac{\tan\alpha}{3}$$

(Ox)

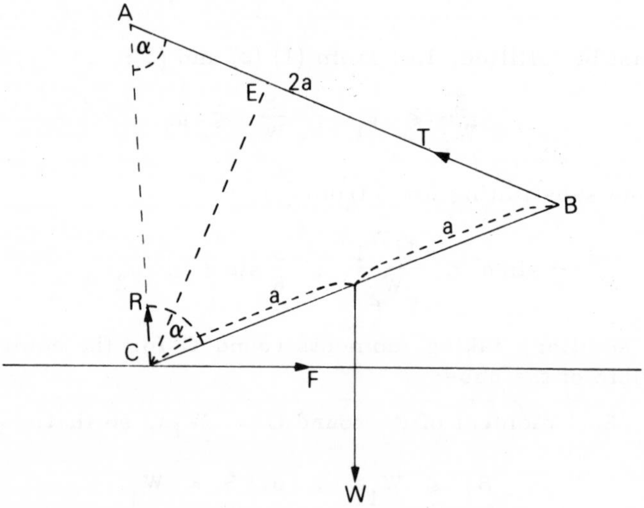

Since AB = BC, $\angle$BCA = $\angle$BAC = a.

Let T be the tension in the string, R and F the normal reaction and force of friction respectively at C acting on the rod. The four forces W, T, F and R acting on the rod are in equilibrium. To find T it is best to take moments round C for the four forces, but first we require the length CE of the perpendicular from C to the line of action of T, which is BA. From triangle CEB:

$$CE = CB \sin CBA = 2a \sin(\pi - 2\alpha) = 2a \sin 2\alpha$$

and so, taking moments round C:

$$Wa \sin \alpha = T \, . \, 2a \sin 2\alpha = 4Ta \sin \alpha \cos \alpha$$

$$\therefore \qquad T = \frac{W}{4 \cos \alpha} \qquad (1)$$

Resolving vertically ahd horizontally:

$$T \cos \alpha + R = W, \text{ and}$$

$$T \sin \alpha = F$$

Consequently:

$$\frac{F}{R} = \frac{T \sin \alpha}{W - T \cos \alpha} = \frac{\frac{W}{4} \tan \alpha}{W - \frac{W}{4}} \text{ from (1)}$$

$$= \frac{\tan \alpha}{3}$$

Now:

$$\frac{F}{R} \leqslant \mu$$

and so μ cannot be less than:

$$\frac{\tan \alpha}{3}$$

6. **A uniform solid hemisphere, weight W and radius r, stands with its circular base in contact with a rough horizontal plane. A uniform rod AB, of length 2a and weight W, is smoothly hinged at A to the horizontal plane, and rests at an angle α with the horizontal against the smooth curved surface of the hemisphere. The rod AB and the centre of the base of the hemisphere both lie in the same vertical plane and C, the point of contact of the rod and the hemisphere, is such that a < AC < 2a. If the coefficient of friction between the hemisphere and the plane is μ, and equilibrium is limiting, show that:**

$$\mu = \frac{a \sin^2 \alpha}{r + a \cos \alpha \sin \alpha}$$

Find the tangent of the angle that the direction of the reaction at A makes with the horizontal.

(AEB).

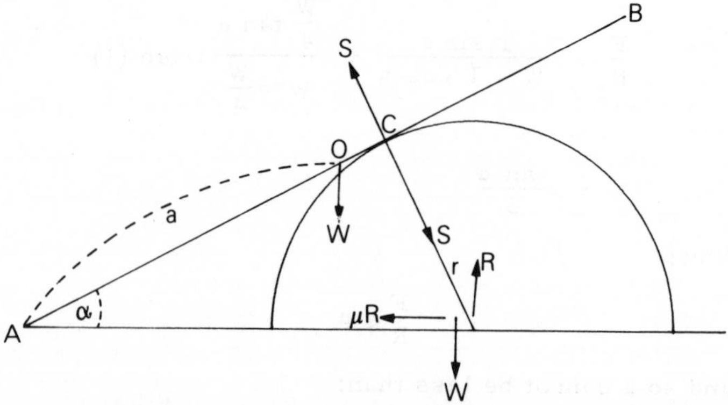

Let O be the centre of the rod, S the action of the hemisphere on the rod at C, which will be normal to it as contact at C is smooth. Let R be the normal reaction of the plane on the hemisphere, then, as equilibrium is limiting, the force of friction will be μR. The rod is in equilibrium under the action of its weight W at O, S, and the action of the hinge at A. Taking moments round A for these:

$$S \cdot r \cot \alpha = W \cdot a \cos \alpha, \text{ i.e.}$$

$$S = \frac{W \cdot a \sin \alpha}{r} \qquad (1)$$

For the equilibrium of the sphere resolve horizontally and vertically:

$$S \sin \alpha = \mu R$$

$$S \cos \alpha + W = R$$

$$\therefore \quad S \sin \alpha = \mu(S \cos \alpha + W), \text{ so that:}$$

$$\mu = \frac{S \sin \alpha}{S \cos \alpha + W}$$

On substituting for S from (1), and reducing, this becomes:

$$\mu = \frac{a \sin^2 \alpha}{r + a \sin \alpha \cos \alpha}$$

Now let the reaction at A be a force P acting in a direction making an angle θ with the horizontal.

Resolving horizontally and vertically for the equilibrium of AB:

$$P \cos \theta = S \sin \alpha$$

$$P \sin \theta + S \cos \alpha = W \qquad \text{Eliminating P:}$$

$$\tan \theta = \frac{W - S \cos \alpha}{S \sin \alpha} = \frac{r - a \sin \alpha \cos \alpha}{a \sin^2 \alpha}$$

on substituting for S from (1).

7. In the diagram, AB is a uniform ladder of length 2a and weight W, and G is the centre of mass of the ladder. The ladder is resting against a fixed cylindrical roller with circular cross-section whose axis is perpendicular to the vertical plane containing AB. The ladder is inclined at an angle $\frac{1}{4}\pi$ to the horizontal. The point of contact X of the ladder with the roller is at a distance $\frac{1}{2}$a from the end A, and the contact at X is smooth. Show that in order that equilibrium be maintained in this position, the coefficient of friction μ at B must not be less than $\frac{1}{2}$.

A man of weight W stands at X and then starts walking slowly up the ladder. Show that if μ = 11/13, the ladder is on the point of slipping when he has moved a distance of $\frac{1}{4}$a.

(Ox)

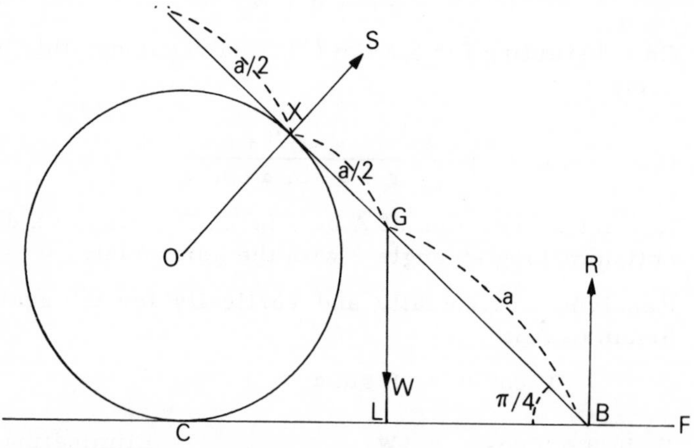

Let O be the centre of the vertical section of the roller against which AB rests, and let the vertical through G meet the line CB in L. Then the ladder is in equilibrium under the action of four forces:

(a) its weight W acting along GL.

(b) S, the reaction of the cylinder at X, which will act along the normal OX, as the contact is smooth.

(c) R, the normal reaction at B.

(d) F, the force of friction at B.

Taking moments round B gives S straightaway:

$$S \cdot \frac{3a}{2} = W \cdot BL = W\frac{a}{\sqrt{2}}$$

so:

$$S = \frac{\sqrt{2} \cdot W}{3} \qquad (1)$$

Resolving horizontally and vertically:

$$S \cos\frac{\pi}{4} = F \qquad (2)$$

$$S \sin\frac{\pi}{4} + R = W \qquad (3)$$

From (1) and (2):

$$F = \frac{W}{3}$$

From (1) and (3):

$$R = \frac{2W}{3}$$

so that:

$$\frac{F}{R} = \frac{1}{2}$$

But, for equilibrium:

$$\frac{F}{R} \leqslant \mu$$

therefore μ must not be less than $\frac{1}{2}$.

When the man has moved a distance a/4, let S, R, F now represent the reaction of the cylinder, the normal reaction and force of friction at B at that instant. Again taking moments round B, for the five forces now acting:

$$W(a + \frac{a}{2} + \frac{a}{4})\frac{1}{\sqrt{2}} + W \cdot \frac{a}{\sqrt{2}} = S \cdot \frac{3a}{2}$$

from which:

$$S = \frac{11W}{12} \cdot \sqrt{2} \qquad (4)$$

Resolving horizontally and vertically, using (4):

$$\frac{11}{12} \cdot W = F \qquad (5)$$

$$\frac{11}{12} \cdot W + R = 2W \qquad (6)$$

and so:

$$R = \frac{13}{12} \cdot W$$

$$\therefore \quad \frac{F}{R} = \frac{11}{13}$$

which, being equal to the coefficient of friction, means that the ladder is on the point of slipping.

8. A framework consists of three light rods AB, BC, CA, of lengths a, a√3, a respectively, smoothly jointed at A, B, C. The framework is suspended freely from A and carries weights 2W at B and W at C. Show that, in the equilibrium position with B below A, the thrust in the rod BC is 2W and that the tensions in the rods AB, CA are 2W√3, W√3 respectively.

(Ox)

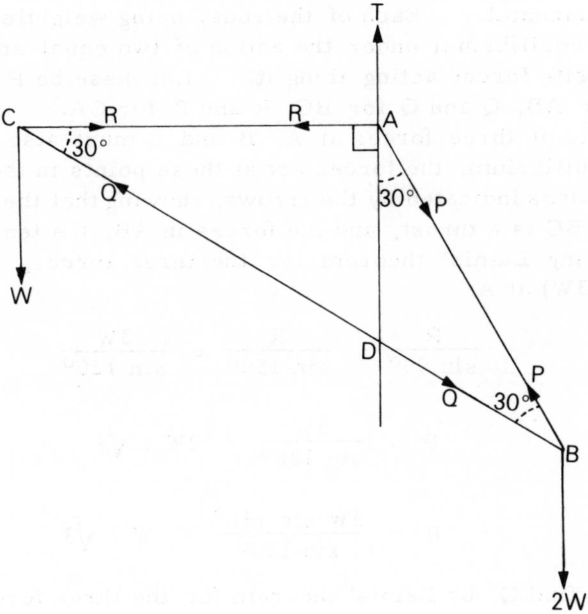

First, the cosine formula for the triangle ABC shows that $\angle BAC = 120^\circ$, so that $\angle ACB = \angle ABC = 30^\circ$. The system of three rods is in equilibrium under the action of three forces, W, 2W and T, the action at A. Hence T must act vertically upwards, also T = 3W. Now let the vertical through A meet BC in D. Then D is on the resultant of the two forces W and 2W, hence:

$$W \cdot CD = 2W \cdot DB$$

It follows that:

$$CD = \frac{2}{3} \cdot CB = \frac{2}{3} \cdot a\sqrt{3} = \frac{2a}{\sqrt{3}} = \frac{a}{\cos 30^\circ}$$

and so $\angle CAD$ must be a right angle, i.e. rod CA is horizontal. Each of the rods, being weightless, is in equilibrium under the action of two equal and opposite forces acting along it. Let these be P and P for AB, Q and Q for BC, R and R for CA. As the sets of three forces at A, B and C must also be in equilibrium, the forces act at these points in the directions indicated by the arrows, showing that the force in BC is a thrust, and the forces in AB, CA tensions. Using Lamis' theorem for the three forces, P, R, T(3W) at A:

$$\frac{P}{\sin 90^\circ} = \frac{R}{\sin 150^\circ} = \frac{3W}{\sin 120^\circ}$$

$$\therefore \quad P = \frac{3W}{\sin 120^\circ} = 2W \cdot \sqrt{3}$$

$$R = \frac{3W \sin 150^\circ}{\sin 120^\circ} = W \cdot \sqrt{3}$$

To find Q, by Lamis' theorem for the three forces at C:

$$\frac{Q}{\sin 90^\circ} = \frac{W}{\sin 150^\circ}$$

so: $Q = 2W$, also got by resolving forces, which gives $Q \cos 60^\circ = W$, and $Q \sin 60^\circ = R$

9. **The ends of a light string are attached to two fixed points A and B at the same horizontal level, so that it sags, the distance AB being 12 m. Five points, C, D, E, F, G, are chosen along the string such that the horizontal projections of AC, CD, DE, EF, FG, GB are all equal. At C and G are attached equal weights 2W, at D and F equal weights 3W, and at E a weight 4W. If E is now 7 m below AB, find:**

(b) the tensions in AC, CD, DE in terms of W.

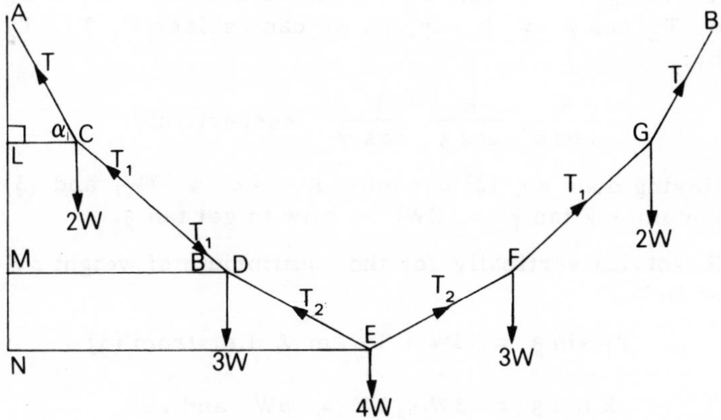

There is symmetry about the vertical through E which will be the lowest point of the string, so it is only necessary to consider the half ACDE. Let α, β, γ be the inclinations of AC, CD, DE respectively to the horizontal, and T, T_1, T_2 the respective tensions in them. Draw CL, DM, EN horizontally to meet the vertical through A in L, M, N. Then, as given, AN = 7 m. Let p be the length of the horizontal projection of each of the 6 portions of string. Then 6p = AB = 12, i.e. p = 2 m. From triangle ALC, since LC = 2, AL = 2 tan α; in the same way LM = 2 tan β, MN = 2 tan γ. Thus:

$$2(\tan\alpha + \tan\beta + \tan\gamma) = 7 \qquad (1)$$

Now considering the equilibrium of the whole system, resolving vertically: $2T\sin\alpha = 4W$, so:

$$T\sin\alpha = 7W \qquad (2)$$

Resolving vertically for the equilibrium of the weight 4W at E, $2T_2\sin\gamma = 4W$, or:

$$T_2 \sin\gamma = 2W \qquad (3)$$

Resolving horizontally for the equilibrium of weight 3W at D, $T_2 \cos\gamma = T_1 \cos\beta$, while, resolving horizontally for the equilibrium of weight 2W at C, $T_1 \cos\beta = T \cos\alpha$. Thus $T \cos\alpha = T_1 \cos\beta = T_2 \cos\gamma = k$ say, so we can replace T, T_1, T_2 by:

$$\frac{k}{\cos\alpha}, \frac{k}{\cos\beta}, \frac{k}{\cos\gamma} \quad \text{respectively.}$$

Having done so, (2) becomes $k \tan\alpha = 7W$, and (3) becomes $k \tan\gamma = 2W$. Now to get $\tan\beta$.

Resolving vertically for the equilibrium of weight 3W at D:

$$T_1 \sin\beta = 3W + T_2 \sin\gamma, \text{ i.e. from (3)}$$

$$k \tan\beta = 3W + 2W = 5W, \text{ and so:}$$

$$\tan\alpha = \frac{7W}{k}\ ; \quad \tan\beta = \frac{5W}{k}\ ; \quad \tan\gamma = \frac{2W}{k}$$

To find k substitute these values for $\tan\alpha$, $\tan\beta$, $\tan\gamma$ in (1). This gives:

$$2 \cdot \frac{14W}{k} = 7\ ; \quad \text{so:}$$

$$k = 4W$$

Accordingly:

$$\tan\alpha = \frac{7W}{4W} = \frac{7}{4}\ ; \quad \tan\beta = \frac{5}{4}\ ; \quad \tan\gamma = \frac{1}{2}$$

From which α, β, and γ can be found from tables.

To find T, T_1, T_2, we have:

$$\cos\alpha = \frac{4}{\sqrt{65}} \; ; \quad \cos\beta = \frac{4}{\sqrt{41}} \; ; \quad \cos\gamma = \frac{2}{\sqrt{5}}$$

so that:

$$T = \frac{k}{\cos\alpha} = 4W \cdot \frac{\sqrt{65}}{4} = W\sqrt{65}$$

Similarly:

$$T_1 = \frac{4W}{\cos\beta} = W\sqrt{41}$$

$$T_2 = 2\sqrt{5} \cdot W$$

10. **Two uniform rods AB and BC, equal in length and each of weight W, are smoothly jointed at B. The end C is connected to the mid-point of AB by a light inelastic string whose length is such that, when it is taut, the triangle ABC is equilateral. The end A is smoothly hinged to a fixed point of a smooth horizontal table, and the rods are in equilibrium in a vertical plane with C resting on the table and the string taut. Find the reaction between C and the table and the tension in the string.**

(Ox)

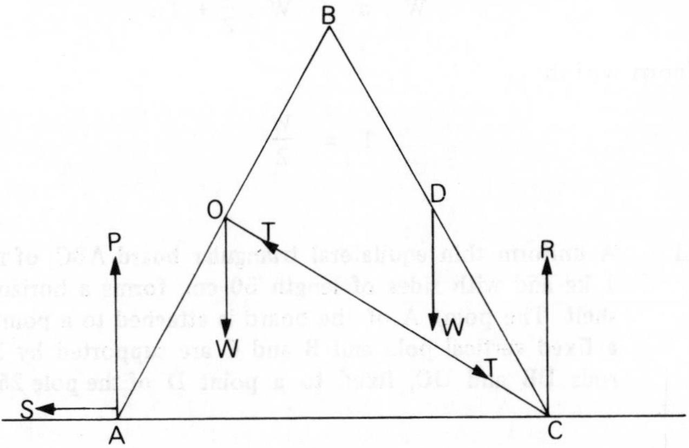

Let AB = BC = 2a. Under the given conditions AC = 2a also.
Let O and D be the middle points of AB and BC, T the tension in the string CO, R the reaction of the smooth table at C, which will thus act vertically upwards, P and S the normal and horizontal components of the reaction of the table at A on the rod AB. The system of two rods is in equilibrium under the action of the external forces which are P, S, R, and the weights W of the rods acting at O and D. When dealing with the whole system we can regard the two weights W as one weight 2W acting through B, midway between O and D.

Resolving horizontally:

$$S = 0$$

Resolving vertically:

$$P + R = 2W \qquad (1)$$

Taking moments around A:

$$2W \cdot a = R \cdot 2a \qquad (2)$$

so:
$$R = W$$

Hence from (1):
$$P = W$$

To find T, take moments round B for the rod AB.

$$W \cdot a = W \cdot \frac{a}{2} + Ta$$

from which:

$$T = \frac{W}{2}$$

11. A uniform thin equilateral triangular board ABC, of mass 1 kg and with sides of length 50 cm, forms a horizontal shelf. The point A of the board is attached to a point on a fixed vertical pole and B and C are supported by light rods DB and DC, fixed to a point D of the pole 25 cm

vertically below A. The mid-point of BC is E. Show that the resultant of the thrusts in DB and DC acts along DE. Prove that angle ADE is 60°, and find, by drawing to scale or by calculation, the magnitude and direction of the reaction at A.

(Ox)

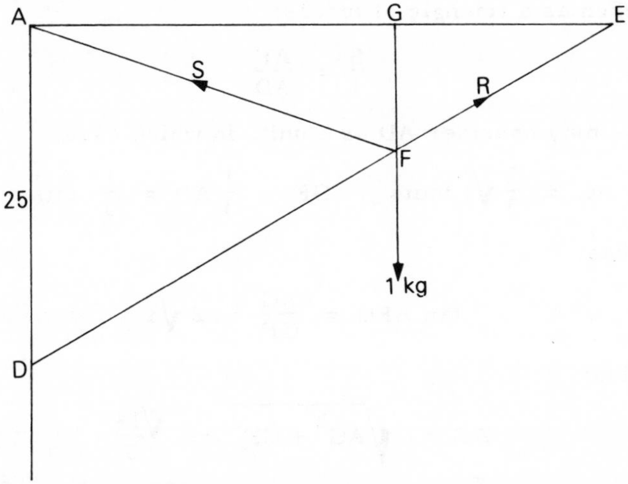

From geometry the rods DB and DC are equal in length, and, from symmetry, the thrusts in them are equal, too. The resultant, R, of these thrusts will therefore bisect the angle BDC, i.e. it will pass through E and so act along DE. Since ABC is an equilateral triangle of side 50 cm:

$$AE = 50 \sin 60^\circ = 25\sqrt{3} \text{ cm}$$

Thus:

$$\tan ADE = \frac{AE}{AD} = \sqrt{3} \ ; \quad \text{so:} \quad \angle ADE = 60^\circ$$

The centre of mass of the board is the centroid of the triangle ABC, which is a point G on AE such that AG = 2/3 AE. Let the vertical through G meet DE in F, and let S kg be the reaction at A. Then the board is in equilibrium under the action of 3 forces, namely its weight (1 kg), R and S. As two of these meet in F, the third, S, must also pass through F, and so will act along FA.

Examining the figure, it can be seen that ADF may be taken as a triangle of forces:

$$\therefore \qquad \frac{S}{1} = \frac{AF}{AD} \qquad (1)$$

We may consider AD as 1 unit, in which case:

$$AG = \frac{2}{3}\sqrt{3} \text{ units} \quad ; \quad GF = \frac{1}{3} AD = \frac{1}{3} \text{ units}$$

Thus:

$$\tan AFG = \frac{AG}{GF} = 2\sqrt{3}$$

Also:

$$AF = \sqrt{AG^2 + GF^2} = \sqrt{\frac{13}{3}}$$

Hence, from (1), the reaction at A is a force of $\sqrt{\frac{13}{3}}$ kg wt, acting at an angle $\tan^{-1}(2\sqrt{3})$ with the upward vertical, in the direction FA.

12. **A uniform block in the form of a cube stands on a plane inclined at an angle a to the horizontal in such a way that four of its edges are parallel to the line of greatest slope. A gradually increasing horizontal force is applied to the uppermost edge of the block at right angles to it and in a vertical plane through the centre of mass of the block, in a direction which would tend to move the block down the plane. If μ (> tan a) is the coefficient of friction between the block and the plane, show that the block will tilt without sliding provided that:**

$$\mu > \frac{2\tan^2 a + \tan a + 1}{\tan^2 a + \tan a + 2}$$

(Ox)

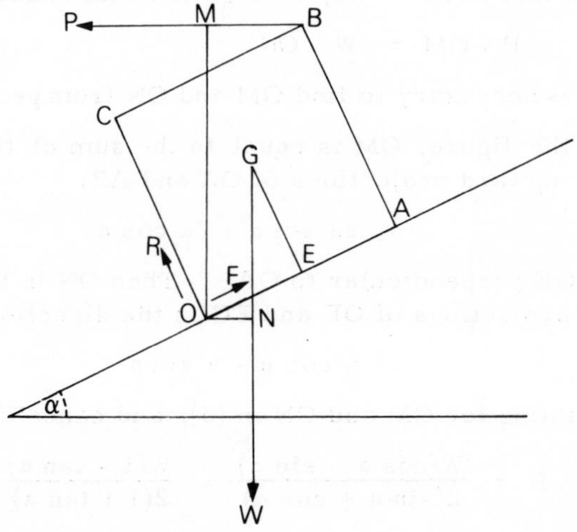

Let W be the weight of the cube, 2a the length of its side, OABC the vertical plane through the centre of mass, G, of the cube. Since $\mu > \tan a$ the block will remain at rest if left to itself. The gradually increasing force is applied at B in the plane OABC.

Suppose that when this force reaches the value P the cube is about to turn round O without sliding. Then, at the instant, the normal reaction, R, of the plane on the cube, and the force of friction, F, both act through O. For this to be possible:

$$\frac{F}{R} < \mu$$

The cube is in equilibrium under the action of the forces W, P, F, R. Resolving along and perpendicular to the plane:

$$F = P\cos\alpha + W\sin\alpha \qquad (1)$$

$$R = W\cos\alpha - P\sin\alpha \qquad (2)$$

To find P take moments round O. Draw ON perpendicular to the line of action of W, and OM perpendicular to that of P. So, taking moments round O:

$$P \cdot OM = W \cdot ON \qquad (3)$$

and it is necessary to find OM and ON from geometry.

From the figure, OM is equal to the sum of the vertically upward projections of OA and AB:

$$= 2a\sin\alpha + 2a\cos\alpha$$

Draw GE perpendicular to OA. Then ON is the sum of the projections of OE and EG in the direction ON:

$$= a\cos\alpha - a\sin\alpha$$

Substituting for OM and ON in (3), and cancelling a:

$$P = \frac{W(\cos\alpha - \sin\alpha)}{2(\sin\alpha + \cos\alpha)} = \frac{W(1 - \tan\alpha)}{2(1 + \tan\alpha)}$$

Hence from (1) and (2):

$$\frac{F}{R} = \frac{P + W\tan\alpha}{W - P\tan\alpha} = \frac{\left(\frac{1 - \tan\alpha}{2(1 + \tan\alpha)} + \tan\alpha\right)}{1 - \frac{\tan\alpha\,(1 - \tan\alpha)}{2(1 + \tan\alpha)})}$$

$$= \frac{2\tan^2 a + \tan a + 1}{\tan^2 a + \tan a + 2}$$

So the block will tilt without sliding if:

$$\mu > \frac{2\tan^2 a + \tan a + 1}{\tan^2 a + \tan a + 2}$$

13. **A uniform ladder of weight W rests with one end on rough horizontal ground and with the other end against a smooth vertical wall. The ladder is at an angle $\tan^{-1} 2$ to the ground and is in a vertical plane perpendicular to the wall. The coefficient of friction between the ladder and the ground is 1/3. Find how far up the ladder a boy of weight 2W can climb without disturbing equilibrium.**
Find also the least horizontal force which must be applied to the foot of the ladder to enable the boy to climb to the top of the ladder without it slipping.

(AEB)

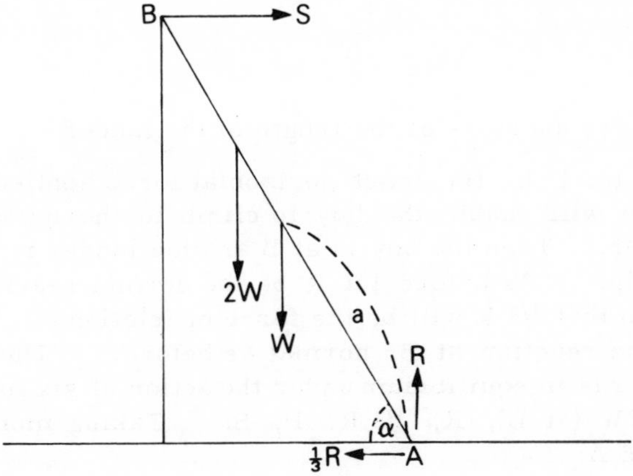

Let AB be the ladder, of length 2a. Denote by α the angle the ladder makes with the ground, so that $\tan \alpha = 2$. Suppose that the boy can climb a distance h up the ladder without disturbing equilibrium, i.e. before it is about to slip. Let R be the normal reaction then at A, so that 1/3 R will be the force of friction. Let S be the reaction of the wall at B, which, since the latter is smooth, will act along the normal to it.

The ladder is now in equilibrium under the action of five forces, W, 2W, R, 1/3 R, S. Resolving vertically:

$$R = W + 2W = 3W \qquad (1)$$

Resolving horizontally:

$$S = 1/3\,R = W \qquad (2)$$

Taking moments round A:

$$S \cdot 2a \sin \alpha = Wa \cos \alpha + 2W \cdot h \cos \alpha$$

i.e. $\quad 2S \cdot a \tan \alpha = Wa + 2Wh$

As $\tan \alpha = 2$, and $S = W$, this becomes, on cancelling W:

$$4a = a + 2h$$

or: $$h = \frac{3a}{2}$$

or three quarters of the length of the ladder.

Now let P be the least horizontal force applied at A which will enable the boy to climb to the top of the ladder. Then the boy is at B and the ladder is about to slip. As before let R be the normal reaction at A, so that 1/3 R will be the force of friction. Let S be the reaction at B, normal as before. Then the ladder is in equilibrium under the action of six forces, W, 2W (at B), R, 1/3 R, P, S. Taking moments round A:

$$S \cdot 2a \sin \alpha = 2W \cdot 2a \cos \alpha + Wa \cos \alpha$$

or: $$2S \tan \alpha = 5W$$

Since $\tan \alpha = 2$:

$$S = \frac{5W}{4}$$

Resolving vertically:

$$R = 2W + W = 3W$$

Resolving horizontally:

$$\frac{1}{3}R + P = S = \frac{5W}{4}$$

$$\therefore \quad W + P = \frac{5W}{4}$$

so that: $$P = \frac{W}{4}$$

14. A thin uniform shelf, of weight W, consists of a rectangular plate ABCD, where BC = a. The shelf is fixed to a vertical wall by a smooth hinge along AB and is supported in the horizontal position by a light rod, smoothly hinged at one end to the underside of the shelf at the mid-point of CD and smoothly hinged at the other end to a point on the wall distance a below the mid-point of AB. Find the thrust in the rod and show, by means of a diagram, the line of action of the force exerted on the shelf by the hinge. Prove that this line of action makes an angle $\frac{1}{4}\pi$ with the vertical.

(Ox)

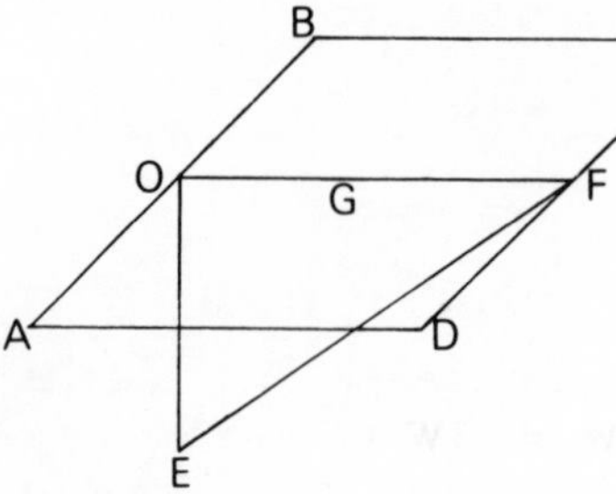

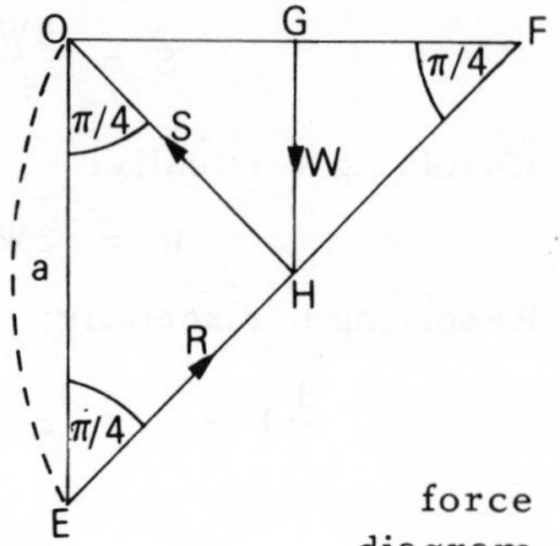

force
diagram

Let O be the mid-point of AB, G centre of shelf, EF the supporting bar. Then EOF is a vertical isosceles right-angled triangle with G the mid-point of OF. Let the vertical through G meet EF in H, then EH = HF. The shelf is in equilibrium under the action of three forces:

(a) its weight W acting along GH.
(b) the action, R, of the light rod along its length EF.
(c) S, the total force exerted on the shelf by the hinge, which, from symmetry, will act through O.

By a theorem concerning three forces in equilibrium, their lines of action must be either parallel or concurrent. Here two of them, R and W, meet at H, so that S must also act through H and so, for equilibrium, must act along HO, as indicated by the arrow. Also $\angle HOE = \frac{1}{4}\pi$, so HO makes an angle of $\frac{1}{4}\pi$ with the vertical. Examining the force diagram, HOE can be taken as a triangle of forces. Therefore:

$$\frac{W}{OE} = \frac{S}{OH} = \frac{R}{EH}$$

and so: $$S = W \cdot \frac{OH}{OE} = \frac{W}{\sqrt{2}} = \frac{W\sqrt{2}}{2}$$

and: $$R = W \cdot \frac{EH}{OE} = \frac{W\sqrt{2}}{2}$$

15. Two uniform ladders AB and AC of equal weights W and equal lengths 2a, are smoothly jointed at A. They rest with the ends B and C on a rough horizontal plane, each ladder making an angle θ with the horizontal. A man of weight W_1 starts to climb the ladder AB. Find whether slipping takes place first at B or C, and, assuming that it does take place, find the distance the man has climbed then, μ being the coefficient of friction at B and C.

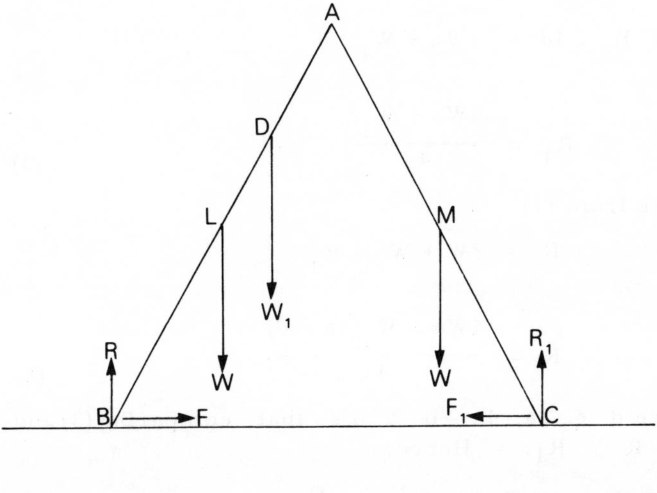

When the man is at D, having climbed a distance BD = d, let R and F be the normal reaction and force of friction at B, R_1 and F_1 the normal reaction and force of friction at C. Let L and M be the middle points of the ladders, through which points their weights, W, act.

Resolving horizontally for the equilibrium of both ladders together:

$$F = F_1, \text{ so we replace } F_1 \text{ by } F \text{ hereafter.}$$

Resolving vertically for both ladders together:

$$R + R_1 = 2W + W_1 \qquad (1)$$

Taking moments about A for the ladder AC:

$$R_1 . 2a \cos\theta = Wa\cos\theta + F . 2a \sin\theta$$

or, cancelling a:

$$2R_1 \cos\theta = W\cos\theta + 2F\sin\theta \qquad (2)$$

Taking moments about B for both ladders:

$$R_1 . 4a\cos\theta = Wa\cos\theta + W_1 . d\cos\theta + W . 3a\cos\theta$$

$$R_1 . 4a = 4Wa + W_1 . d$$

i.e.
$$R_1 = \frac{4Wa + W_1 d}{4a} \qquad (3)$$

Thus from (1):

$$R = 2W + W_1 - R_1$$

so:
$$R = \frac{4Wa + W_1(4a - d)}{4a} \qquad (4)$$

Since $d < 2a$, $4a - d > d$ so that, comparing (3) and (4), $R > R_1$. Hence:

$$\frac{F}{R} < \frac{F}{R_1}$$

which shows that slipping will take place first at C.

From (2):

$$F = \frac{(2R_1 - W)\cot\theta}{2}$$

$$= \frac{(2Wa + W_1 d)\cot\theta}{4a}$$

and so: $$\frac{F}{R_1} = \frac{(2Wa + W_1 d)\cot\theta}{4Wa + W_1 d}$$

When: $$\frac{F}{R_1} = \mu$$

the ladder is about to slip at C, and the man has climbed a distance d given by:

$$\mu = \frac{(2Wa + W_1 d)\cot\theta}{4Wa + W_1 d}$$

from which: $$d = \frac{2Wa(2\mu - \cot\theta)}{W_1(\cot\theta - \mu)}$$

For d to be positive, $2\mu > \cot\theta > \mu$ is necessary.

16. Three uniform rods AB, BC, CD of lengths 2a, a, 2a, and weights 2W, W, 2W respectively, are smoothly jointed at B and C. The rods AB, CD rest on two small smooth horizontal pegs at the same level and at a distance 6a/5 apart. The whole system lies symmetrically in equilibrium in a vertical plane with BC horizontal and AB, CD inclined at an angle θ to the vertical. Show that $\theta = 1/6\ \pi$. Show also that the reaction at the hinge B makes an angle $\tan^{-1} 5\sqrt{3}$ with the vertical.

(Ox)

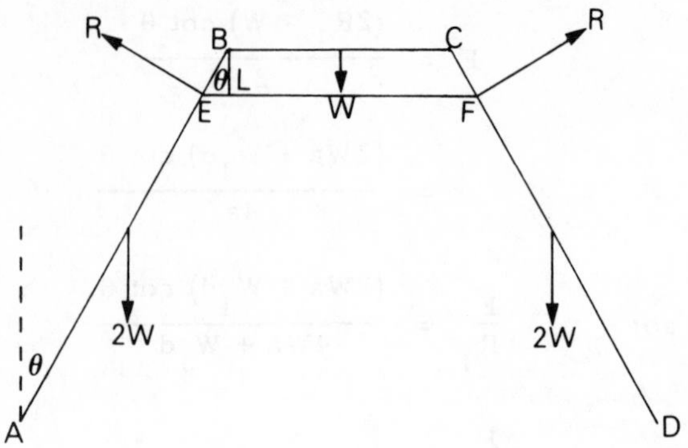

Let E and F be the pegs, R the reaction on the rod at each peg, equal because of symmetry. These reactions will be normal to the rods, the pegs being smooth. Consider first the equilibrium of the system formed by the three rods. Resolving vertically:

$$2R \sin\theta = 5W, \text{ so } R = \frac{5W}{2\sin\theta} \qquad (1)$$

As a preliminary to taking moments about B for the rod AB, the length BE has to be found. Draw BL perpendicular to EF. Since EF = 6a/5, from symmetry:

$$EL = \tfrac{1}{2}\left(\frac{6a}{5} - a\right) = \frac{a}{10}$$

$$\therefore \quad BE = \frac{EL}{\sin\theta} = \frac{a}{10\sin\theta}$$

Now taking moments about B for the rod AB:

$$2W \,.\, a\sin\theta = R \,.\, BE = \frac{5W}{2\sin\theta} \,.\, BE, \text{ from (1)}$$

i.e. $$2W \cdot a \sin \theta = \frac{5W}{2 \sin \theta} \cdot \frac{a}{10 \sin \theta}$$

from which: $$\sin^3 \theta = \frac{1}{8}$$

and: $$\sin \theta = \tfrac{1}{2}$$

or: $$\theta = 30^\circ$$

Hence from (1): $$R = 5W$$

Now consider the rod BC. From symmetry the forces acting on it at B and C will be equal. Let each be a force S at an angle θ with the vertical.

Resolving vertically for the equilibrium of BC:

$$2S \cos \theta = W \qquad (2)$$

But an equal force S acts at B on AB in the opposite direction. Resolving horizontally for the equilibrium of rod AB:

$$S \sin \theta = R \cos 30^\circ = 5W \cos 30^\circ \qquad (3)$$

Division of (3) by (2) gives:

$$\tan \theta = 5\sqrt{3}$$

17. A particle of unit mass is moving along a straight line and is attracted to a fixed point O in the line by a force equal to 2x pdl, where x ft is the distance of the particle from O at time t sec. In addition the motion of the particle is resisted by a force equal to 2v pdl, where v ft/sec is the velocity. Write down a differential equation to determine the motion of the particle and hence show that $x = Ce^{-t} \cos (t + a)$, where C and a are constants.

If, initially, $x = 3$ and $v = 0$, find the values of C and α, and the value of v when $t = \pi/2$.

(AEB)

When particle is at a distance x ft from O its velocity and acceleration are:

$$\frac{dx}{dt} \quad \text{and} \quad \frac{d^2x}{dt^2}$$

The total resisting force in pdl is:

$$2x + 2\frac{dx}{dt}$$

Hence the equation of motion of the particle is:

$$\frac{d^2x}{dt^2} = -2x - 2\frac{dx}{dt}$$

i.e. $$\frac{d^2x}{dt^2} + 2\frac{dx}{dt} + 2x = 0 \qquad (1)$$

A differential equation of this type is solved by letting $x = e^{kt}$, where k is to be determined. Then:

$$\frac{dx}{dt} = k \cdot e^{kt}$$

$$\frac{d^2x}{dt^2} = k^2 \cdot e^{kt}$$

After substitution (1) becomes:

$$e^{kt}(k^2 + 2k + 2) = 0$$

This will be satisfied if $k^2 + 2k + 2 = 0$, whose roots are:

$$-1 \pm \sqrt{-1}, \text{ i.e. } -1 \pm i$$

And so:

$$e^{(-1+i)t} \text{ and } e^{(-1-i)t}$$

are particular solutions of (1), thus its general solution will be:

$$Ae^{-(1-i)t} + Be^{-(1+i)t}$$

in which A and B are arbitrary constants. Since $e^{\pm it} = \cos t \pm i \sin t$, this may be written as:

$$e^{-t}(A(\cos t + i \sin t) + B(\cos t - i \sin t))$$

or, changing the constants, as:

$$e^{-t}(E \cos t + F \sin t)$$

This last factor, $E \cos t + F \sin t$ can be put in form $C \cos(t + \alpha)$. Therefore the general solution of (1) is:

$$x = Ce^{-t} \cos(t + \alpha) \qquad (2)$$

where C and α are constants.

Differentiating with respect to t:

$$\frac{dx}{dt} = -Ce^{-t}(\cos(t + \alpha) + \sin(t + \alpha)) \qquad (3)$$

Initially $x = 3$ and $v = 0$, Therefore, letting $t = 0$ in (2):

$$3 = C \cdot \cos \alpha$$

Similarly from (3):

$$0 = -C(\cos \alpha + \sin \alpha)$$

Thus $C \cdot \sin \alpha = -C \cdot \cos \alpha = -3$, from which:

$$C = \pm 3\sqrt{2}.$$

There are now two cases:

(a) if $C = +3\sqrt{2}$:

$$\cos a = \frac{1}{\sqrt{2}}, \quad \sin a = \frac{-1}{\sqrt{2}}, \quad a = -\frac{\pi}{4}$$

(b) if $C = -3\sqrt{2}$:

$$\cos a = \frac{-1}{\sqrt{2}}, \quad \sin a = \frac{1}{\sqrt{2}}, \quad a = \frac{3\pi}{4}$$

When $t = \frac{\pi}{2}$, from (3), in case (a):

$$\frac{dx}{dt} = v = -3\sqrt{2} \,.\, e^{-\pi/2}\left(\cos\frac{\pi}{4} + \sin\frac{\pi}{4}\right) = -6e^{-\pi/2}$$

When $t = \frac{\pi}{2}$, in case (b):

$$v = -(-3\sqrt{2})e^{-\pi/2}\left(\cos\frac{5\pi}{4} + \sin\frac{-5\pi}{4}\right) = -6e^{-\pi/2}$$

as in case (a).

18. **Two small spheres, A and B, of masses m and 2m respectively lie at rest on a smooth horizontal floor. The spheres are situated in a line perpendicular to a vertical wall, with sphere B nearer to the wall than sphere A. The sphere A impinges directly on sphere B with velocity u; after rebounding from the wall, sphere B is brought to rest by its second impact with sphere A. If the coefficient of restitution between sphere A and sphere B is 1/3:**

(a) show that the coefficient of restitution between sphere B and the wall is 1/5.

(b) find the final velocity of sphere A.

(AEB)

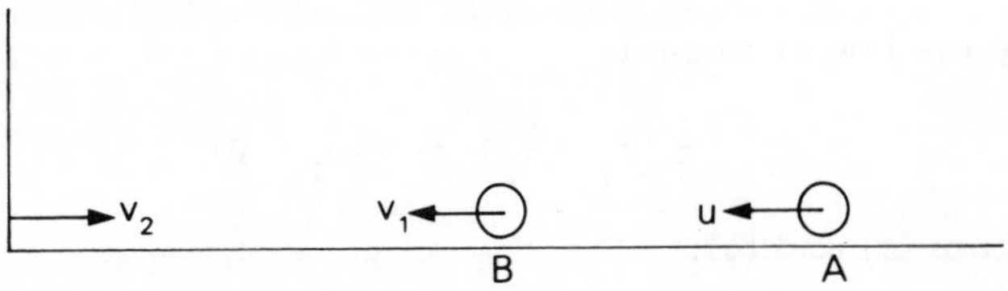

Let u_1 and v_1 be the velocities of A and B respectively, after their first collision, in the direction AB. Then from the law of conservation of momentum:

$$mu = mu_1 + 2mv_1$$

or: $$u_1 + 2v_1 = u \qquad (1)$$

From the law of impact:

$$v_1 - u_1 = \frac{1}{3} \cdot u \qquad (2)$$

Solving equations (1) and (2) for u_1 and v_1:

$$u_1 = \frac{u}{9} \; ; \quad v_1 = \frac{4u}{9}$$

Let e be the coefficient of restitution between sphere B and the wall. Then after its impact with the wall B will rebound along BA with a velocity v_2 given by:

$$v_2 = ev_1 = \frac{4eu}{9} \qquad (3)$$

Let u_2 be the velocity of A after its second impact with sphere B, which is thereby brought to rest. For the second impact of the spheres, from the law of conservation of momentum:

$$mu_1 - 2mv_2 = mu_2$$

i.e. cancelling m, from (3):

$$\frac{u}{9} - \frac{8eu}{9} = u_2 \qquad (4)$$

By the law of impact:

$$\frac{1}{3}(u_1 + v_2) = (-u_2 - 0) \qquad (5)$$

From (4) and (5):

$$u\left(\frac{1}{9} - \frac{8e}{9}\right) = \frac{-u}{3}\left(\frac{1}{9} + \frac{4e}{9}\right)$$

and so, cancelling u: $e = \frac{1}{5}$

On substituting in (4): $u_2 = \frac{-u}{15}$

i.e. A has a final velocity $\frac{u}{15}$ along BA.

19. A smooth ring P of mass m is threaded on to a fixed horizontal wire OX and a second smooth ring Q of equal mass is threaded on a fixed wire OY where Y is vertically below O. The rings are connected by a light inextensible string of length 2a. Initially P and Q are at rest with P at a distance $a\sqrt{3}$ from O, and Q at O. Q is then released. Prove that the initial velocity of P is $\frac{1}{4}\sqrt{(6ga)}$, and find the velocity of Q immediately after the string has become taut.
If θ is the inclination of PQ to the horizontal, prove that when both P and Q are in motion:

$$\left(\frac{d\theta}{dt}\right)^2 = \frac{g}{8a}(8\sin\theta - 1)$$

and find the tension in the string at this time.

(Ox)

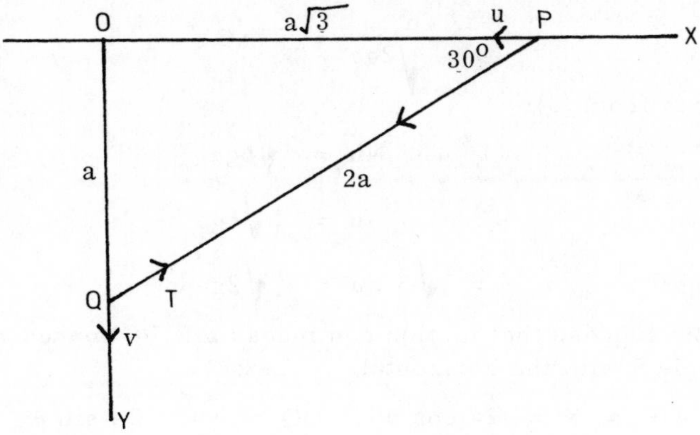

Since the length of the string is 2a and initially OP = $a\sqrt{3}$, it follows that when Q has fallen a vertical distance a, thus acquiring a velocity $\sqrt{2ga}$, the string becomes taut, and the $\angle OPQ = 30^o$. Let u and v be the initial velocities of P and Q respectively (in the directions indicated by the arrows) after this has taken place, and T the impulsive tension in the string. Then since impulse = change in momentum:

$$mu = T \cos 30^o \qquad (1)$$

$$m(v - \sqrt{2ga}) = -T \sin 30^o \qquad (2)$$

Also since the string is inextensible, the velocities of P and Q in the direction PQ must be equal, and so:

$$u \cos 30^o = v \sin 30^o$$

i.e.

$$v = \sqrt{3} \, . \, u \qquad (3)$$

Division of (1) by (2) gives:

$$\frac{u}{v - \sqrt{2ga}} = -\sqrt{3}$$

Thus from (3):

$$4u = \sqrt{6ga}$$

$$u = \tfrac{1}{4}\sqrt{6ga}$$

also: $$v = \sqrt{3} \,.\, u = \tfrac{3}{4}\sqrt{2ga}$$

Now suppose that motion continues until PQ makes an angle θ with the horizontal. Let:

$$OP = x = 2a \cos\theta \quad ; \quad OQ = y = 2a \sin\theta$$

Let U and V denote the present velocities of P and Q, u and v denoting the initial velocities already found. Then by the principle of the conservation of energy, the change in the kinetic energy of the system equals the work done by the external forces:

$$= mg(2a \sin\theta - a)$$

Thus:

$$\tfrac{1}{2}mU^2 + \tfrac{1}{2}mV^2 - \tfrac{1}{2}mu^2 - \tfrac{1}{2}mv^2 = mga(2 \sin\theta - 1) \quad (4)$$

Now:

$$U = \frac{dx}{dt} = \frac{d(2a\cos\theta)}{dt} = -2a \sin\theta \,.\, \frac{d\theta}{dt}$$

It is convenient, for short, to denote:

$$\frac{d\theta}{dt} \quad \text{and} \quad \frac{d^2\theta}{dt^2}$$

by $\dot{\theta}$ and $\ddot{\theta}$ respectively in a question like this, and so:

$$U = -2a \sin\theta \,.\, \dot{\theta}$$

Similarly: $$V = \frac{dy}{dt} = 2a \cos\theta \,.\, \dot{\theta}$$

Therefore, substituting in (4) for U, V, u, v:

$$2a^2 \dot{\theta}^2 - \tfrac{1}{2} \cdot \frac{9}{16} \cdot 2ga - \tfrac{1}{2} \cdot \frac{1}{16} \cdot 6ga = ga(2 \sin\theta - 1)$$

which, after simplification, gives:

$$\dot{\theta}^2 = \frac{g}{8a} \cdot (8 \sin\theta - 1) \qquad (5)$$

To find the tension, T, in thesstring at this time, differentiate (5) with respect to t, which gives:

$$2\dot{\theta} \cdot \ddot{\theta} = \frac{g}{a} \cos\theta \cdot \dot{\theta}$$

or, cancelling $\dot{\theta}$:

$$a\ddot{\theta} = \frac{g}{2} \cos\theta \qquad (6)$$

From the equation of motion of P, whose acceleration along OP is:

$$\frac{dU}{dt} = -2a(\sin\theta \cdot \ddot{\theta} + \cos\theta \cdot \dot{\theta}^2)$$

The force acting on P in the direction PO is T cos θ. Hence:

$$m \cdot 2a(\sin\theta \cdot \ddot{\theta} + \cos\theta \cdot \dot{\theta}^2) = T \cos\theta$$

or: $$T = 2m(a\ddot{\theta} \cdot \tan\theta + a\dot{\theta}^2)$$

Substituting for $a\dot{\theta}^2$ and $a\ddot{\theta}$ from (5) and (6), this becomes:

$$T = 2m\left(\frac{g}{2} \sin\theta + \frac{g}{8}(8 \sin\theta - 1)\right)$$

$$= \frac{mg}{4}(12 \sin\theta - 1)$$

20. **Two trains start from station A and complete the 5 mile journey to station B in 8 minutes. The first train stops at station B having accelerated steadily to its top speed of 60 mile/h and been brought to rest by a constant retardation. Find the distance this train travels at 60 mile/h. The second train accelerates uniformly for the first 2 miles and then travels with constant speed, passing through station B at this top speed. Find the top speed and the acceleration of the second train.**

(AEB)

Suppose that the times of acceleration and retardation of the first train are t min, and t_1 min respectively, so that the train travels at constant speed of 60 mile/h for $(8 - t - t_1)$ minutes. A speed of 60 mile/h is equal to a speed of 1 mile/min.

Distance travelled during acceleration = average speed x time = $\frac{1}{2}t$ miles.

Similarly distance travelled during retardation = $\frac{1}{2}t_1$ miles.

Distance travelled at constant speed = $1 . (8 - t - t_1)$ miles. Hence:

$$\tfrac{1}{2}(t + t_1) + 1 . (8 - t - t_1) = 5$$

and so:
$$t + t_1 = 6$$

$$8 - t - t_1 = 2$$

Thus the distance the train travels at 60 mile/h is 2 miles.

Now let V mile/min be the top speed of the second train, and t min be the time during which it accelerates. Then distance travelled during acceleration = average speed x time, so:

$$\frac{Vt}{2} = 2 \text{ miles} \quad ; \quad Vt = 4 \tag{1}$$

Since train now travels 3 miles in (8 - t) min at speed V:

$$V(8 - t) = 3 \tag{2}$$

Thus from (1) and (2):

$$8V - 4 = 3$$

$$V = \frac{7}{8} \text{ mile/min} = 52{\cdot}5 \text{ mile/h}$$

The constant acceleration, f, is found from the formula:

$$V^2 = 2fs$$

$$f = \frac{V^2}{2s} = \frac{49}{4 \times 64} = \frac{49}{256} \text{ mile/min}^2$$

$$= \frac{539}{1920} \text{ ft/s}^2$$

21. **A triangular framework ABC is formed by three rods AB, BC and CA joined at their ends. The rods AB and AC are uniform and of equal mass, but are unequal in length. The rod BC is light and of length a, and the angle BAC is a right angle. The framework can turn freely about a horizontal axis through A at right angles to its plane. Prove that the period of small oscillations about the stable equilibrium position is the same as for a simple pendulum of length 2a/3.**

(Ox)

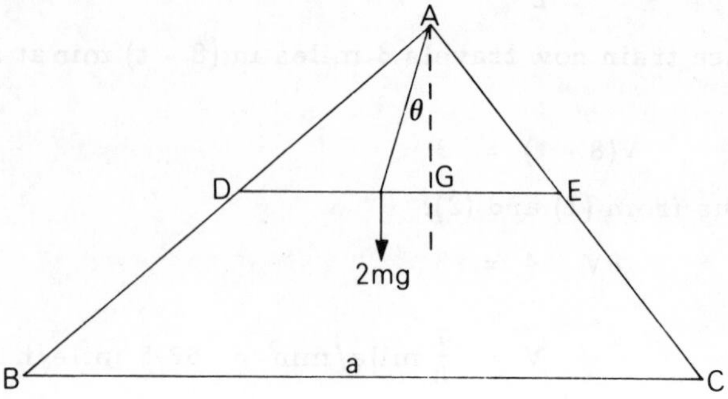

Let m be the mass of each of the rods AB and AC, D and E their centres of mass, which will be their mid-points. The light rod BC is there merely to preserve rigidity. The centre of mass of the system will thus be at G, the mid-point of DE. Denote by L the horizontal axis through A at right angles to the plane of the rods. In the position of stable equilibrium AG is vertical. If I is the moment of inertia of the system about L, the period of small oscillations is derived from the usual formula, which, as the total mass is 2m, here becomes:

$$I \frac{d^2\theta}{dt^2} = -2mg\ AG\ \theta \qquad (1)$$

where θ is the small angle through which the framework is rotated. It is now necessary to find I and AG. First, since D and E are the middle points of AB and AC:

$$DE = \frac{BC}{2} = \frac{a}{2}$$

Also, since DAE is a right angle:

$$AG = DG = \frac{DE}{2} = \frac{a}{4}$$

The moment of inertia of AB about L is, by the usual formula for a rod:

$$m \cdot \frac{AB^2}{3}$$

Similarly that of AC about L is:

$$m \cdot \frac{AC^2}{3}$$

Therefore, by addition:

$$I = \frac{m}{3}(AB^2 + AC^2) = m \cdot \frac{BC^2}{3} = m \cdot \frac{a^2}{3}$$

On substituting for I and AG in (1) it becomes:

$$m \cdot \frac{a^2}{3} \cdot \frac{d^2\theta}{dt^2} = -2mg \cdot \frac{a}{4} \cdot \theta$$

or:

$$\frac{d^2\theta}{dt^2} = -\frac{3g}{2a} \cdot \theta$$

This represents simple harmonic motion whose period:

$$2\pi\sqrt{\frac{2a}{3g}}$$

is the same as that of a simple pendulum of length 2a/3.

22. **A thin uniform plate of mass m has the form of a rhombus ABCD, of side 2a and with $\angle$ BAD = 1/3 π. Show that the moment of inertia of the plate about the line normal to its plane and passing through its centre of mass is 2/3 ma^2.**

The plate is suspended with its plane vertical from a fixed horizontal axis passing through B and is free to swing sbout this axis. Show that the period of small oscillations is the same as that of a simple pendulum of length 5 a/3.

(Ox)

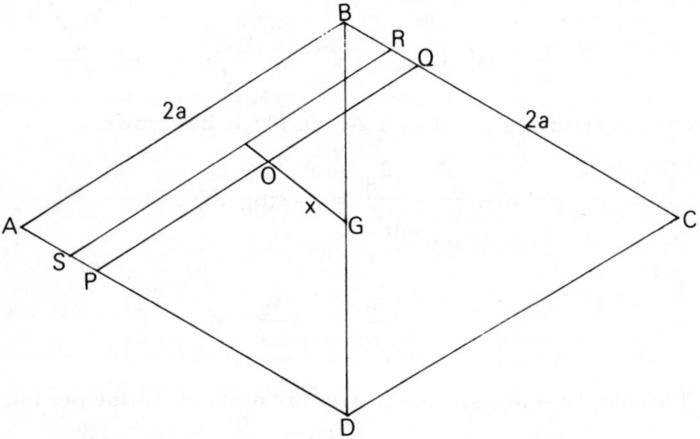

Since $\angle$BAD = $\pi/3$, the triangles BAD and BCD will be equilateral. Let G be the point of intersection of the diagonals, which will also be the centre of mass of the plate. Let L be the axis through G normal to the plane of the plate, and denote by ρ the mass per unit area of the plate. The area of the latter being $4a^2 \sin 60^\circ$, it follows that:

$$4\rho \cdot a^2 \sin 60^\circ = m \qquad (1)$$

Divide the plate into thin strips parallel to one side, say to AB, and consider one of them, PQRS. Let O be the centre of PQ, and let GO = x, OV = δx. Then this strip may be regarded as a thin uniform rod of length 2a. Its mass, δm, = ρ x area = $\rho \cdot 2a \cdot \delta x \sin 60^\circ$. Hence the moment of inertia of this strip about an axis through its centre O parallel to L will be:

$$\delta m \cdot \frac{a^2}{3} = \rho \cdot 2a \cdot \delta x \sin 60^\circ \cdot \frac{a^2}{3}$$

Therefore by the theorem of parallel axes (see note on example 1) the moment of inertia of this strip about L will be:

$$\rho \cdot 2a \cdot \delta x \sin 60^\circ \cdot \frac{a^2}{3} + \rho \cdot 2a \cdot \delta x \sin 60^\circ \cdot x^2$$

Hence by summation for all the strips, the moment of inertia of the plate about L:

$$= 2\int_0^a \left\{ \rho \cdot 2a \sin 60^\circ \left(\frac{a^2}{3} + x^2\right) \right\} \delta x$$

$$= 4\rho a \sin 60^\circ \left(\frac{a^3}{3} + \frac{a^3}{3}\right)$$

$$= \frac{2}{3} ma^2 \text{ from (1)}$$

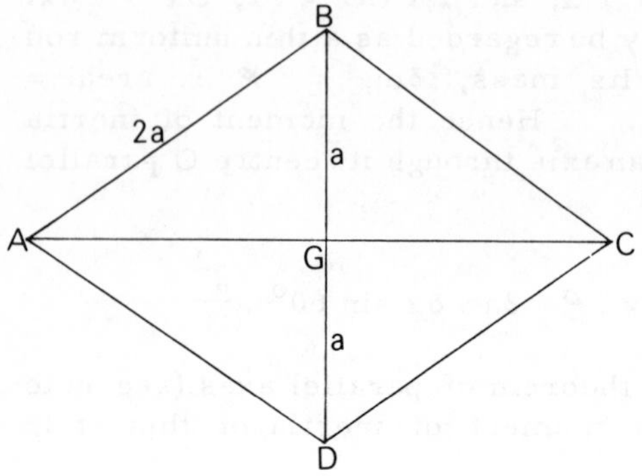

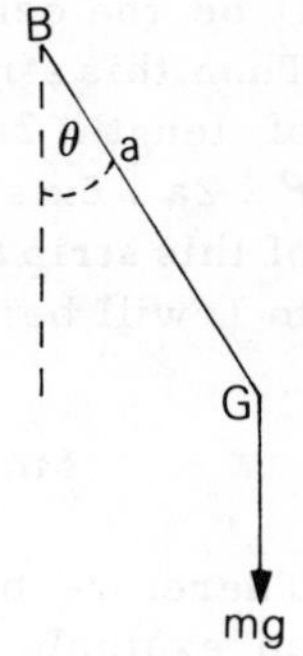

When the plate is suspended from a fixed horizontal axis through B, which will be perpendicular to its plane, let I be the moment of inertia of the plate about this axis. Now let the plate be rotated through a small angle θ about B, so that its moment of momentum about the fixed axis will be:

$$I \cdot \frac{d\theta}{dt}$$

at the instant. Since the rate of change of the moment of momentum about a fixed axis equals the moment of the external forces about that axis:

$$I \cdot \frac{d^2\theta}{dt^2} = -mga \sin\theta = -mga\theta, \text{ as } \theta \text{ is small.}$$

But, again using the theorem of parallel axes:

$$I = \text{moment of inertia about L} + m \cdot BG^2$$

$$= \frac{2}{3} ma^2 + ma^2 = \frac{5}{3} ma^2$$

$$\therefore \quad \frac{5}{3}ma^2 \cdot \frac{d^2\theta}{dt^2} = -mga\theta$$

or:

$$\frac{d^2\theta}{dt^2} = -\frac{3g}{5a}\theta$$

which shows that the plate swings in simple harmonic motion of period:

$$2\pi\sqrt{\frac{5a}{3g}}$$

This is the same as that of a simple pendulum of length l where l = 5a/3.

23. **A flywheel has a horizontal axis of radius r. The system has a mass M and radius of gyration k about its axis and rotates without friction.**
A string is wound around the axle and carries a mass m hanging freely. If the system is released from rest, prove that the acceleration of the mass m is:

$$g\left(\frac{mr^2}{Mk^2 + mr^2}\right)$$

If the string slips off the axle after the mass m has descended a distance 8r, find the magnitude of the constant retarding couple which is necessary to bring the flywheel to rest in n more revolutions.

(AEB)

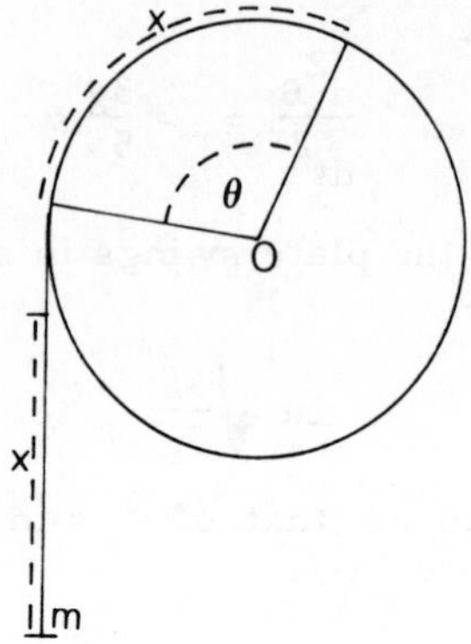

Suppose the flywheel turns through an angle θ in time t, and that the mass m has descended a distance x. Then since there is no friction:

$$x = r\theta \tag{1}$$

For short use $\dot{x}$, $\ddot{x}$ etc to denote:

$$\frac{dx}{dt}, \frac{d^2x}{dt^2} \text{ etc}$$

Then from (1): $\dot{x} = r\dot{\theta}$; $\ddot{x} = r\ddot{\theta}$.

Let T be the tension in the string after time t. Then the equation of motion for mass m is:

$$m\ddot{x} = mg - T$$

i.e.
$$mr\ddot{\theta} = mg - T \tag{2}$$

Also, for the flywheel, the rate of change of its moment of momentum about its axis = moment of external forces about the axis:

i.e. $$Mk^2\ddot{\theta} = Tr \qquad (3)$$

To eliminate T, multiply (2) by r and add to (3).

Hence: $$(Mk^2 + mr^2)\,\ddot{\theta} = mgr$$

and so: $$\ddot{\theta} = \frac{mgr}{Mk^2 + mr^2}$$

Thus the acceleration f of the mass m $= \ddot{x} = r\ddot{\theta}$:

$$= g\left(\frac{mr^2}{Mk^2 + mr^2}\right)$$

After the mass m has descended a distance 8r its velocity v is given by $v^2 = 2 \cdot 8r \cdot f = 16rf$.

$$v^2 = 16rg\left(\frac{mr^2}{Mk^2 + mr^2}\right)$$

If ω is the angular velocity of the flywheel at that instant:

$$v = r\omega$$

$$\therefore \quad r^2 \cdot \omega^2 = 16rg\left(\frac{mr^2}{Mk^2 + mr^2}\right)$$

$$\omega^2 = \frac{16mgr}{Mk^2 + mr^2}$$

and so the kinetic energy of the flywheel then:

$$= \tfrac{1}{2}Mk^2\omega^2 = \frac{8M\,mk^2\,gr}{Mk^2 + mr^2}$$

Let C be the constant retarding couple which will bring the flywheel to rest in n more revolutions. The work done by a couple C in a turn through an angle β radians is C . β. Consequently, from the principle of energy:

$$C \cdot 2n\pi = \frac{8M\,mk^2\,gr}{Mk^2 + mr^2}$$

$$\therefore \qquad C = \frac{4M\,mk^2\,gr}{n\pi\,(Mk^2 + mr^2)}$$

24. **A particle is projected horizontally with speed u across a smooth horizontal plane from a point O in the plane. The particle is subjected to a retardation of magnitude k times the speed of the particle. Find the distance of the particle from O and also its speed at time t after projection.**
Another particle is projected vertically upwards with speed u. In addition to the retardation due to gravity, this particle is also subjected to a retardation of k times its speed. Find the time this particle takes to reach its greatest height.

(AEB)

Suppose that after a time t the particle is at a distance x from O, and has a speed v. As the plane is smooth, particle has no retardation except kv, hence:

$$\frac{dv}{dt} = -kv$$

Integrating, log v = -kt + C, where C is a constant. Initially, when t = 0, v = u, so that C = log u.

$$\therefore \qquad \log v = -kt + \log u$$

which may be written as:

$$v = u \cdot e^{-kt}$$

i.e. $$\frac{dx}{dt} = u \cdot e^{-kt}$$

Integrating again:

$$x = -\frac{u}{k} \cdot e^{-kt} + A$$

where A is a constant.

When $t = 0$, $x = 0$, so $A = \frac{u}{k}$.

$$\therefore \quad x = \frac{u}{k}\left(1 - e^{-kt}\right)$$

In the case of the particle projected vertically upwards, if v be its velocity after time t:

$$\frac{dv}{dt} = -g - kv$$

or: $$\frac{dv}{g + kv} = -dt$$

Integration gives:

$$\frac{1}{k} \log (g + kv) = -t + D$$

where D is a constant.

When $t = 0$, $v = u$, so that $D = \frac{1}{k}(g + ku)$.

Hence: $$t = \frac{1}{k} \log \left(\frac{g + ku}{g + kv}\right)$$

The greatest height is attained when $v = 0$, which will thus be after a time:

$$T = \frac{1}{k} \log \left(\frac{g + ku}{g}\right)$$

25. A small circular hoop, of radius a and centre O, fixed in a vertical plane, has a small smooth ring of mass m threaded on it and held at a point level with O. A light elastic string, of natural length 2a and modulus of elasticity 3 mg, is fixed at one end to the lowest point A of the hoop. From A the string passes vertically upwards, over a small smooth peg fixed at the highest point B of the hoop, and then straight across to the ring, to which its other end is attached. The ring is then projected with velocity $\sqrt{ag}$ upwards. At a subsequent time, t, measured from the instant of projection, it is at P, where the angle PBA is θ ($\pi/4 < \theta < \pi/2$). Prove that the speed of the ring is then $\sqrt{(2ag)}\sin\theta$, and show that the time taken for the ring to reach P is:

$$\sqrt{\left(\frac{2a}{g}\right)}\log_e\left\{(\sqrt{2}+1)\tan\tfrac{1}{2}\theta\right\}$$

(Ox)

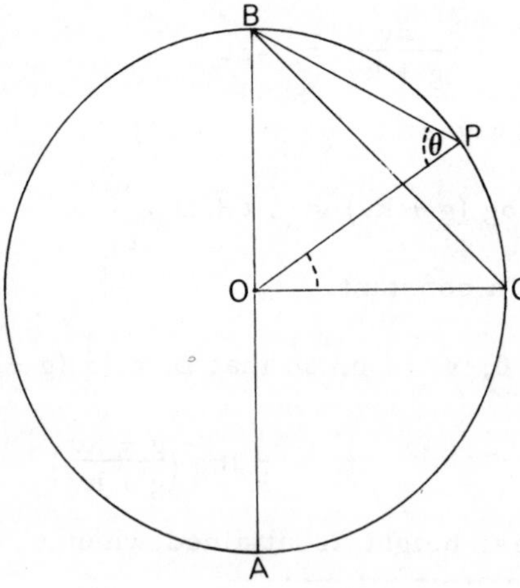

Let the ring be at C initially, so that $\angle BOC = \pi/2$, $\angle OBC = \pi/4$. When the ring is at P let its velocity be v, then its loss of kinetic energy is:

$$\tfrac{1}{2} mag - \tfrac{1}{2} mv^2 \qquad (1)$$

We now need this formula. The work done by an elastic string of modulus λ and natural length ℓ when its extension is shortened from x_1 to x_2 is equal to:

$$\text{(the average tension)} \times (x_1 - x_2)$$

$$= \frac{\lambda}{2\ell}(x_1 + x_2)(x_1 - x_2) = \frac{\lambda}{2l}(x_1^2 - x_2^2)$$

Apply this to the present case, in which $\lambda = 3mg$; $\ell = 2a$; $x_1 = BC = \sqrt{2}a$; $x_2 = BP = 2a \cos\theta$. It follows that the work done by the elastic string while the ring moves from C to P is:

$$\frac{3mg}{4a}(2a^2 - 4a^2 \cos^2\theta) \qquad (2)$$

Let $\angle COP = \psi$. Then since $\angle COP$ at centre of circle $= 2\angle CBP$ at circumference, we have:

$$\psi = 2(\theta - \frac{\pi}{4}) = 2\theta - \frac{\pi}{2} \qquad (3)$$

The work done in raising the ring from C to P is, therefore:

$$mg \,.\, a \sin\psi = mg \,.\, a \sin(2\theta - \frac{\pi}{2}) = -mg \,.\, a \cos 2\theta \qquad (4)$$

Now applying the principle of the conservation of energy, it follows from (1), (2) and (4) that:

$$\tfrac{1}{2} mga - \tfrac{1}{2} mv^2 + \frac{3mg}{4a}(2a^2 - 4a^2\cos^2\theta) =$$

$$-mg \,.\, a \cos 2\theta$$

or: $$ga - v^2 + 3ga - 6ga \cos^2\theta = -2ga \cos 2\theta$$

Replacing:

$$\cos^2\theta \text{ by } 1 - \sin^2\theta$$

and: $$\cos 2\theta \text{ by } 1 - 2\sin^2\theta$$

$$v^2 = 4ga - 6ga(1 - \sin^2\theta) + 2ga(1 - 2\sin^2\theta) =$$

$$2ga\sin^2\theta$$

$$\therefore \quad v = \sqrt{(2ga)}\sin\theta$$

Now: $$v = a\frac{d\psi}{d\theta} = 2a\frac{d\theta}{dt}$$ from (3) and so:

$$2a\frac{d\theta}{dt} = \sqrt{(2ga)}\sin\theta$$

i.e. $$\sqrt{(\frac{2a}{g})}\frac{d\theta}{\sin\theta} = dt$$

Integrating this equation:

$$\sqrt{(\frac{2a}{g})}\log_e\tan\frac{\theta}{2} = t + C$$

where C is a constant to be found.

When $t = 0$, $\theta = \frac{\pi}{4}$; so that $C = \sqrt{(\frac{2a}{g})}\log\tan\frac{\pi}{8}$

Therefore, substituting for C gives:

$$\sqrt{(\frac{2a}{g})}\log_e\left(\frac{\tan\frac{\theta}{2}}{\tan\frac{\pi}{8}}\right) = t$$

Now: $\cot \frac{\pi}{8} = \sqrt{2} + 1$

And so: $t = \sqrt{\left(\frac{2a}{g}\right)} \log_e \left\{(\sqrt{2} + 1) \tan \frac{1}{2} \theta\right\}$

26. **A particle is projected with velocity V, at an angle of elevation α to the horizontal, from a point on a plane inclined at angle β ($< \alpha$) to the horizontal. The path of the particle is in a vertical plane through a line of greatest slope of the plane. If R_1 and R_2 are the respective maximum ranges when the particle is fired up the plane and down the plane, show that:**

(a) $$R_1 = \frac{V^2}{g(1 + \sin \beta)}$$

(b) $$\sin \beta = \frac{R_2 - R_1}{R_2 + R_1}$$

(AEB)

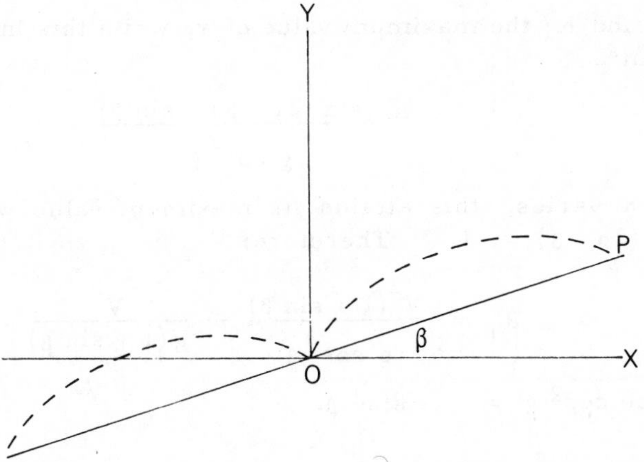

Take the horizontal and vertical lines through O, the point from which the particle is projected, as the x and y axes. Then the equation of the path described by the particle is found, in the usual way, to be:

$$y = x \tan \alpha - \frac{gx^2}{2V^2 \cos^2 \alpha} \qquad (1)$$

If this meets the plane again at a point P, let the range OP = r. Then the co-ordinates of P, which are r cos β and r sin β, satisfy (1), and so:

$$r \sin \beta = r \cos \beta \tan \alpha - \frac{gr^2 \cos^2 \beta}{2V^2 \cos^2 \alpha}$$

Dividing across by r, this gives:

$$r = \frac{2V^2 \cos^2 \alpha (\cos \beta \tan \alpha - \sin \beta)}{g \cos^2 \beta}$$

which reduces to:

$$r = \frac{2V^2 \cos \alpha \sin (\alpha - \beta)}{g \cos^2 \beta}$$

To find R_1 the maximum value of r, write this in the form:

$$V^2 \frac{(\sin (2\alpha - \beta) - \sin \beta)}{g \cos^2 \beta}$$

As α varies, this attains its maximum value when sin (2α - β) = 1. Therefore:

$$R_1 = \frac{V^2(1 - \sin \beta)}{g \cos^2 \beta} = \frac{V^2}{g(1 + \sin \beta)}$$

since $\cos^2 \beta = 1 - \sin^2 \beta$.

To find R_2, it is only necessary to replace β by $-\beta$ in the expression for R_1, and so:

$$R_2 = \frac{V^2}{g(1 - \sin\beta)}$$

$$\therefore \quad \frac{R_2 - R_1}{R_2 + R_1} = \frac{\dfrac{1}{1 - \sin\beta} - \dfrac{1}{1 + \sin\beta}}{\dfrac{1}{1 - \sin\beta} + \dfrac{1}{1 + \sin\beta}} = \sin\beta$$

27. **A particle starts from the origin O and moves in a horizontal plane with a constant acceleration $(-2i + j)$ cm/s^2, i and j being unit vectors in the directions of the co-ordinate axes Ox and Oy respectively, the unit of distance being 1 cm on each axis. The initial velocity of the particle is $(9i - 4j)$ cm/s. By considering the resolved parts of the motion of the particle in the directions Ox and Oy, or otherwise, show that the position vector of the particle after 5 seconds is:**

$$20i - \left(\frac{15}{2}\right)j$$

Find the position vector of the particle after t seconds, and hence show that the particle moves along the curve whose equation is:

$$x^2 + 4xy + 4y^2 - 8x - 18y = 0$$

(AEB)

Consider the resolved part of the motion of the particle in the direction Ox. Its initial velocity is 9, its acceleration -2. Hence from the formula:

$$S = ut + \tfrac{1}{2} f t^2$$

the distance moved in that direction in 5 seconds is:

$$9 \times 5 + \frac{25}{2}(-2) = 20 \text{ cm}$$

Now consider the resolved part of the motion of the particle in the direction Oy. Its initial velocity is -4, its acceleration 1. And so the distance moved in that direction in 5 seconds is:

$$5(-4) + \frac{25}{2} = -\frac{15}{2} \text{ cm}$$

Combining these two results, the position vector of the particle after 5 seconds is:

$$20i - (\frac{15}{2})j$$

Similarly after t seconds the particle has moved a distance in the direction Ox given by:

$$x = 9t - 2 \cdot \frac{t^2}{2} = (9t - t^2) \text{ cm} \qquad (1)$$

and a distance in the direction Oy given by:

$$y = (-4t + \frac{t^2}{2}) \text{ cm} \qquad (2)$$

The position vector of the particle after t seconds is, therefore:

$$(9t - t^2)i + (-4t + \frac{t^2}{2})j$$

To find the equation of the curve along which the particle moves, eliminate t between the equations (1) and (2). Multiply (2) by 2, and add the result to (1).

$$\therefore \qquad 2y + x = t$$

Substitution of this for t in (1) gives, on reduction:

$$x^2 + 4xy + 4y^2 - 8x - 18y = 0$$

28. **Two small rings, each of mass m, are connected by a light inelastic string of length a. One ring, A, is free to slide on a rough horizontal wire, the coefficient of friction between the ring and the wire being μ, whilst the other ring, B, hangs freely. Initially A is at rest with the string vertical and B is moving with a horizontal velocity $\sqrt{(2ga)}$ in the vertical plane containing the wire. Show that the ring A begins to move at some stage if $\mu < \frac{3}{4}$.**

(Ox)

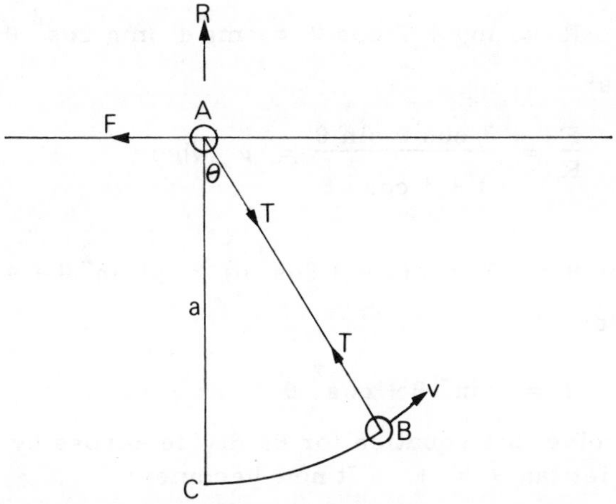

Suppose that, initially, ring B was at C, and that after a time t the string AB makes an angle θ with AC. Let v be the velocity of B at that instant, T the tension in AB, assuming that the ring A remains at rest, R and F the normal reaction and force of friction acting on ring A. Then the equation of motion of B along the radius BA is:

$$m\frac{v^2}{a} = T - mg\cos\theta \qquad (1)$$

Also by the principle of conservation of energy:

$$\tfrac{1}{2}m \cdot 2ga - \tfrac{1}{2}mv^2 = mg(a - a\cos\theta) \qquad (2)$$

From (2): $v^2 = 2ga\cos\theta$, and so, from (1):

$$T = 2mg\cos\theta + mg\cos\theta = 3mg\cos\theta$$

Now consider the equilibrium of A. Resolving along and perpendicular to the wire:

$$F = T\sin\theta = 3mg\cos\theta \cdot \sin\theta$$

$$R = mg + T\cos\theta = mg + 3mg\cos^2\theta$$

Thus:

$$\frac{F}{R} = \frac{3\cos\theta\sin\theta}{1 + 3\cos^2\theta} = \mu \text{ when:}$$

$$3\cos\theta\sin\theta = \mu(1 + 3\cos^2\theta) = \mu(\sin^2\theta + 4\cos^2\theta)$$

since:

$$1 = \sin^2\theta + \cos^2\theta$$

To solve this equation for θ, divide across by $\cos^2\theta$, and let $\tan\theta = t$. It now becomes:

$$3t = \mu(t^2 + 4) \;; \text{ or: } \mu t^2 - 3t + 4\mu = 0 \qquad (3)$$

The roots of this quadratic for t are real if $9 \geqslant 16\mu^2$, i.e. if:

$$3 \geqslant 4\mu$$

Therefore if:

$$\mu < \tfrac{3}{4}$$

the ring A begins to move at some stage, i.e. when t satisfies the quadratic (3).

29. **Two identical spheres each of mass m collide when both are moving with speed u on a smooth horizontal table in the directions shown in the figure. On impact the centres of the spheres are on the line XY. The coefficient of restitution between the spheres is $\frac{1}{2}$. After impact A moves at right angles to XY. Find, after impact, the speed of B, and the angle that the path of B makes with XY. Show that 30 % of the kinetic energy of the spheres is lost in the collision.**

(AEB)

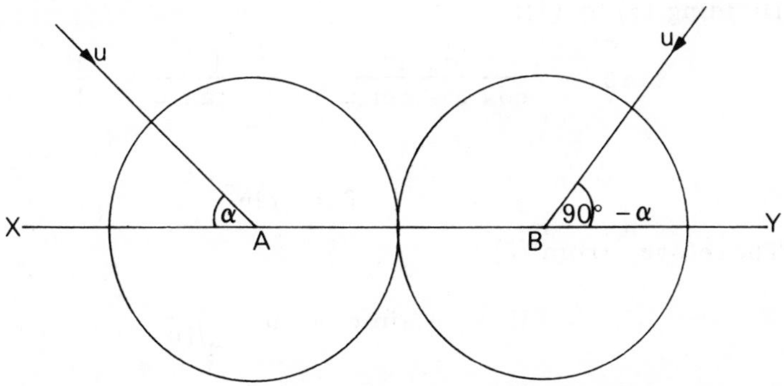

Let the velocity of A after impact be U, and that of B after impact be V at an angle β with XY.

From the law of conservation of momentum along XY:

$$mu \cos \alpha - mu \sin \alpha = mV \cos \beta$$

i.e. $$u(\cos \alpha - \sin \alpha) = V \cos \beta \qquad (1)$$

Also, since impact is smooth:

$$u \sin \alpha = U, \text{ from A} \qquad (2)$$

$$u \cos \alpha = V \sin \beta, \text{ from B} \qquad (3)$$

From the law of impact:

$$\tfrac{1}{2}(u \cos \alpha + u \sin \alpha) = V \cos \beta \qquad (4)$$

Equations (1) and (4) show that:

$$u(\cos \alpha - \sin \alpha) = \tfrac{1}{2}(u \cos \alpha + u \sin \alpha)$$

which give:

$$\tan \alpha = \frac{1}{3}$$

so:

$$\alpha = \tan^{-1} \frac{1}{3}$$

Dividing (3) by (1):

$$\tan \beta = \frac{\cos \alpha}{\cos \alpha - \sin \alpha} = \frac{1}{1 - \tan \alpha} = \frac{3}{2}$$

$$\beta = \tan^{-1} \frac{3}{2}$$

Therefore, from (2):

$$U = u \sin \alpha = u \cdot \frac{1}{\sqrt{10}}$$

From (3):

$$u \cdot \frac{3}{\sqrt{10}} = V \cdot \frac{3}{\sqrt{13}}$$

and so:

$$V = u\sqrt{\frac{13}{10}}$$

Thus the loss of kinetic energy in the collision:

$$= \tfrac{1}{2}mu^2 + \tfrac{1}{2}mu^2 - \tfrac{1}{2}mU^2 - \tfrac{1}{2}mV^2$$

$$= mu^2 \left(1 - \frac{1}{2} \cdot \frac{1}{10} - \frac{1}{2} \cdot \frac{13}{10}\right) = mu^2 \cdot \frac{3}{10}$$

which is a loss of 30%.

30. **A parachutist of mass m falls freely until his parachute opens. When it is open, he experiences an upward resistance where v is his speed and k is a positive constant. Prove that, after time t from the opening of his parachute:**

$$m\left(\frac{dv}{dt}\right) = mg - kv$$

Prove also that, irrespective of his speed when he opens his parachute, his speed approaches a limiting value mg/k, providing that he falls for a sufficiently long time.

The parachutist falls from rest freely under gravity for a time m/2k and then opens his parachute. Prove that the total distance he has fallen when his velocity is:

$$\frac{3mg}{4k} \quad \text{is:} \quad \frac{m^2g}{8k^2}\,(8 \log_e 2 - 1)$$

(Ox)

After time t from the opening of the parachute, the upward resistance kv, and his weight mg downwards are the only forces acting on the parachutist. Since his acceleration is:

$$\frac{dv}{dt}$$

his equation of motion is:

$$m\,\frac{dv}{dt} = mg - kv$$

This may be written:

$$\frac{mdv}{mg - kv} = dt$$

Integrating:

$$\frac{-m}{k}\log_e(mg - kv) = t + C$$

where C is a constant.

To find C, suppose that when $t = 0$, $v = V$. Then:

$$C = \frac{-m}{k}\log(mg - kV)$$

and so: $$t = \frac{m}{k}\log\left(\frac{mg - kV}{mg - kv}\right)$$

Let t tend to infinity, then so does:

$$\frac{mg - kV}{mg - kv}$$

being always positive, so that $mg - kv$ tends to 0. And so v approaches the limiting value mg/k. Since:

$$\frac{dv}{dt} = v\frac{dv}{ds}$$

the equation of motion may also be written in the form:

$$mv\frac{dv}{ds} = mg - kv$$

i.e. $$\frac{mv\,dv}{mg - kv} = ds \qquad (1)$$

To integrate this, let $mg - kv = u$, so that v is replaced by:

$$\frac{mg - u}{k} \quad \text{and dv by:} \quad \frac{-du}{k}$$

This transforms (1) into:

$$\frac{m}{k^2}\left(1 - \frac{mg}{u}\right)du = ds$$

Integration gives:

$$\frac{mu}{k^2} - \frac{m^2 g}{k^2} \log_e u = S + C \qquad (2)$$

where C is a constant. To find C, after time m/2k:

$$v = \frac{gm}{2k} \qquad \text{and } S = \frac{g}{2}\left(\frac{m}{2k}\right)^2$$

Since u = mg - kv, it follows that:

$$u = \frac{mg}{2} \text{ after time } \frac{m}{2k}$$

Substituting for S and u in (2):

$$\frac{m^2 g}{2k^2} - \frac{m^2 g}{k^2} \log\left(\frac{mg}{2}\right) = \frac{gm^2}{8k^2} + C$$

from which:

$$C = \frac{3}{8}\frac{m^2 g}{k^2} - \frac{m^2 g}{k^2}\log\left(\frac{mg}{2}\right)$$

With this value for C, (2) becomes:

$$\frac{mu}{k^2} = \frac{m^2 g}{k^2}\log\left(\frac{2u}{mg}\right) + \frac{3}{8}\frac{m^2 g}{k^2} + S \qquad (3)$$

Now when $v = \frac{3mg}{4k}$; $u = mg - kv = \frac{mg}{4}$.

Substituting for u in (3) the corresponding value of S is seen to be:

$$S = \frac{m^2 g}{4k^2} + \frac{m^2 g}{k^2}\log 2 - \frac{3}{8}\frac{m^2 g}{k^2}$$

$$= \frac{m^2 g}{8k^2}(8 \log_e 2 - 1)$$

31. **A particle of mass m hangs in equilibrium, being attached to the lower end of an elastic string of natural length a and modulus λ, whose upper end is fixed at a point O. It is then given a vertical displacement downwards and allowed to oscillate. Find the greatest distance through which the particle subsequently rises.**

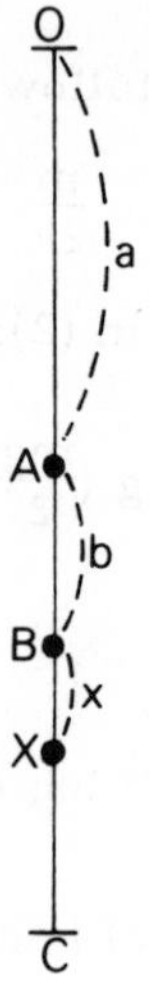

Measure lengths vertically downwards from O. Let OA = a so that OA is the natural length of the string. When the particle takes up an equilibrium position at B, denote AB, the extension of the string, by b. Then, by Hooke's law the tension T = $\lambda b/a$. Thus, since the particle is in equilibrium, mg - T = 0, and so:

$$mg = \frac{\lambda b}{a} \tag{1}$$

Now suppose that the particle is given a vertical displacement downwards, to a point X where BX = x say.

The acceleration is:

$$\frac{d^2x}{dt^2}$$

downwards, or $\ddot{x}$

The forces acting on the particle are mg downwards and the tension in the string which is now:

$$\frac{\lambda(b + x)}{a}$$

Hence, forming its equation of motion:

$$m\ddot{x} = mg - \frac{\lambda(b + x)}{a} = \frac{-\lambda x}{a} \text{ from (1)} \qquad (2)$$

Thus: $$\ddot{x} = \frac{-\lambda x}{am} = -\frac{g}{b} . x \text{ from (1).}$$

Let: $\omega^2 = \frac{g}{b}$, so that this becomes:

$$\ddot{x} = -\omega^2 x \qquad (3)$$

This represents simple harmonic motion whose centre is B. Since the particle starts from rest at X, the amplitude will be x. If $x \leqslant b$, the particle will perform complete oscillations, and the greatest distance it rises subsequently above X will be 2x, the string never becoming slack. Now suppose the particle be given a vertically downward displacement to C, where $BC = c > b$, and then allowed to oscillate. Let its depth below B be x at time t. Equation (3) still holds good, its general solution being:

$$x = A \cos \omega t + B \sin \omega t$$

where A and B are constants. Here $x \equiv c$, when $t = 0$, so $A = c$. Also $\dot{x} = 0$, when $t = 0$, thus $B = 0$. As a result:

$$x = c \cos \omega t \qquad (4)$$

The velocity, v, is from (4):

$$v = \dot{x} = -\omega c \sin \omega t$$

from which:

$$v^2 = \omega^2 (c^2 - x^2) \qquad (5)$$

When $x = -b$, the particle reaches A, with a velocity V given by:

$$V^2 = \omega^2 (c^2 - b^2)$$

The string has now its natural length, and subsequently becomes slack. The particle will now move under gravity only and rise through a height h above A given by:

$$h = \frac{V^2}{2g} = \frac{\omega^2 (c^2 - b^2)}{2g}$$

$$= \frac{c^2 - b^2}{2b}$$

on substituting for ω^2.

Therefore the total height the particle rises above C will be:

$$c + b + h = \frac{(b + c)^2}{2b}$$

32. A particle of mass λm hangs at rest at the end of a light inextensible string of length ℓ which has its other end attached at a fixed point A. A uniform rod, of length 2ℓ and mass m, has one end smoothly hinged at A. The rod is held at rest in a horizontal position and released to rotate and to strike the particle which comes momentarily to rest when level with A. Find the angular speed of the rod just before impact and the speed of the particle just after impact.

If the coefficient of restitution between the rod and the particle is $\frac{1}{2}$, find λ and the angular speed of the rod just after impact.

(AEB)

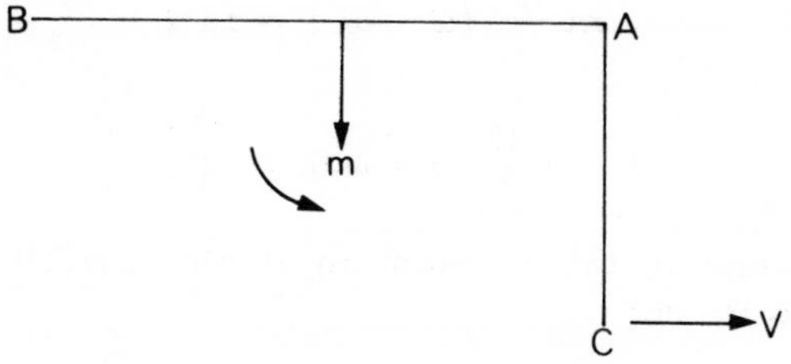

Let AB be the initial position of the rod, C that of the particle. Just before impact, let ω be the angular speed of the rod, about A, and let V be the speed of the particle just after impact. Since the moment of inertia of the rod about its axis of rotation at A is:

$$m \cdot \frac{4\ell^2}{3}$$

by the principle of energy:

$$\tfrac{1}{2}m \cdot \frac{4\ell^2}{3} \omega^2 = mg\ell \; ; \quad \text{i.e. } \omega^2 = \frac{3g}{2\ell}$$

so that: $$\omega = \sqrt{\frac{3g}{2\ell}}$$

Now apply the principle of energy to the motion of the particle. From the given conditions:

$$\tfrac{1}{2}\lambda m \cdot V^2 = \lambda m \cdot g\ell \quad \text{or:} \quad V = \sqrt{2g\ell}$$

Let ω_1 be the angular speed of the rod just after impact and P the mutual impulse between rod and particle. Then since change in the moment of momentum of the rod about the fixed axis through A equals the moment of the impulse about the axis:

$$m\frac{4l^2}{3}(\omega - \omega_1) = Pl$$

or:

$$m\frac{4l}{3}(\omega - \omega_1) = P \qquad (1)$$

But change in the momentum of the particle is also equal to P, so:

$$\lambda m \, . \, V = P$$

i.e.

$$\lambda m \, . \, \sqrt{2gl} = P \qquad (2)$$

As the coefficient of restitution between rod and particle is $\frac{1}{2}$, by the law of impact:

$$V - l\omega_1 = \tfrac{1}{2}l\omega$$

so that:

$$\sqrt{2gl} - l\omega_1 = \tfrac{1}{2}l\sqrt{\frac{3}{2}\frac{g}{l}}$$

from which follows:

$$\omega_1 = \left(\frac{4 - \sqrt{3}}{2\sqrt{2}}\right)\sqrt{\frac{g}{l}}$$

Now to find λ. From (1) and (2):

$$\lambda = \frac{4l(\omega - \omega_1)}{3\sqrt{2gl}}$$

$$= \frac{2\sqrt{2}}{3}\sqrt{\frac{l}{g}}\left(\sqrt{\frac{3}{2}} - \frac{4 - \sqrt{3}}{2\sqrt{2}}\right)\sqrt{\frac{g}{l}}$$

$$= \frac{1}{3}(3\sqrt{3} - 4)$$

33. **At time t a particle of mass m, moving in a straight line, has speed v and is at a distance x from its positon when t = 0. The variables v, x and t are connected by the relation:**

v = at — bx, where a and b are positive constants, and v = 0 when t = 0.

By differentiation and intergration prove that the force acting on the particle at time t is:

$$ma \, . \, e^{-bt}$$

Find x in terms of t and the given constants; and show when t is large, x is approximately:

$$\frac{a(t - \frac{1}{b})}{b} \qquad \textbf{(Ox)}$$

Differentiate the given equation v = at - bx with respect to t, remembering that:

$$\frac{dx}{dt} = v$$

This gives:

$$\frac{dv}{dt} = a - bv \qquad (1)$$

which may be written as:

$$\frac{dv}{a - bv} = dt$$

Integrate both sides:

$$-\frac{1}{b} \log (a - bv) = t + C$$

where C is a constant to be found. When:

$$t = 0, \; v = 0, \; \text{so } C = -\frac{1}{b} \log (a)$$

Hence:

$$\log\left(\frac{a - bv}{a}\right) = -bt \quad \text{or:} \quad a - bv = a \,.\, e^{-bt} \qquad (2)$$

Now the force acting on the particle at time t is:

$$m\frac{dv}{dt}$$

$$= m(a - bv) \qquad \text{from (1)}$$

$$= ma \,.\, e^{-bt} \qquad \text{from (2)}$$

To find x, from (2), $bv = a - ae^{-bt}$.

Integrate both sides with respect to t:

$$\therefore \qquad bx = at + \frac{a}{b}e^{-bt} + A$$

where A is a constant.

When $t = 0$, $x = 0$, so that $A = -\frac{a}{b}$. Accordingly:

$$bx = at + \frac{a}{b}e^{-bt} - \frac{a}{b}$$

or:

$$x = \frac{a}{b}\left(t - \frac{1}{b}\right) + \frac{a}{b^2} \,.\, e^{-bt}$$

When t is large, e^{bt} is very large in comparison so that e^{-bt} is very small, and:

$$\frac{a}{b^2} \,.\, e^{-bt}$$

may be ignored. Therefore x is approximately:

$$\frac{a}{b}\left(t - \frac{1}{b}\right)$$

34. A particle A, of mass m, rests on a rough horizontal table, the coefficient of friction between the particle and the table being μ. A light elastic string, of modulus of elasticity mg and natural length a, is connected at one end to the particle A, passes over a small smooth pulley fixed at the edge of the table and is attached at the other end to another particle B, also of mass m. The particle B is held vertically below the pulley, with the string just taut, and is then released. Prove that, if $\mu < 2$, the particle A begins to move when B has descended a distance μa, and the time taken for this descent is:

$$\sqrt{\left(\frac{a}{g}\right)} \cos^{-1} (1 - \mu) \qquad \text{(Ox)}$$

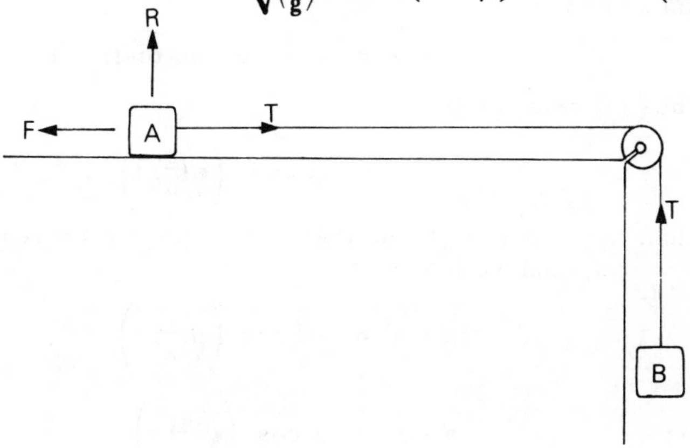

After particle B has descended a distance x the tension T in the string will be:

$$mg \cdot \frac{x}{a}$$

by Hooke's law of elasticity, assuming that A remains at rest. The equation of motion for B is, therefore:

$$m\ddot{x} = mg - mg\frac{x}{a} \quad \text{or:} \quad \ddot{x} = -\frac{g}{a}(x - a)$$

If we replace x - a by y this becomes:

$$\ddot{y} = -\frac{g}{a}y$$

which is the equation occurring in simple harmonic motion, its solution being:

$$y = A\cos\left(\sqrt{\left(\frac{g}{a}\right)}\,t + B\right) \qquad (1)$$

where A and B are constants to be determined from the initial conditions. Differentiating (1):

$$\dot{y} = -A\sqrt{\left(\frac{g}{a}\right)}\sin\left(\sqrt{\left(\frac{g}{a}\right)}\,t + B\right)$$

But when t = 0:

$$\dot{y} = \dot{x} = 0 \quad \text{so that:} \quad B = 0$$

Thus (1) reduces to:

$$y = A\cos\left(\sqrt{\left(\frac{g}{a}\right)}\,t\right)$$

when t = 0, x = 0, so that y = -a. As a result A = -a, and we have:

$$x - a = -a\cos\left(\sqrt{\left(\frac{g}{a}\right)}\,t\right)$$

or:

$$a - x = a\cos\left(\sqrt{\left(\frac{g}{a}\right)}\,t\right) \qquad (2)$$

Assuming that A does not slip, let F be the force of friction acting on it. Then:

$$F = T = mg\,.\,\frac{x}{a}$$

while, if R is the normal reaction on A, R = mg, and so:

$$\frac{F}{R} = \frac{x}{a}$$

Slipping occurs when:

$$\frac{F}{R} = \mu$$

i.e. when: $$x = \mu a$$

From (2) this takes place after the time given by:

$$a - a\mu = a \cos\left(\sqrt{\left(\frac{g}{a}\right)}\, t\right)$$

or: $$t = \sqrt{\left(\frac{a}{g}\right)} \cos^{-1}(1 - \mu) \qquad (3)$$

Since $\cos\theta \geqslant -1$, we must have $1 - \mu \geqslant -1$ or:

$$\mu \leqslant 2$$

Therefore if $\mu < 2$, particle A begins to move when B has descended a distance μa, the time t taken being given by (3).

35. **The figure shows a straight smooth horizontal groove AB in which each of the perfectly elastic particles P and Q can move freely. The vertical line OY is perpendicular to AB and passes through a fixed point M of the groove. Initially P and Q are at rest at distances a and a/2 respectively from OY. The masses of P and Q are m and 2m respectively. An attractive force, λ . PO, where λ is a positive constant, acts continuously on particle P which starts to move towards Q which is unaffected throughout by this attraction. Show that:**

(a) **P moves along the groove with simple harmonic motion about M as centre.**

(b) **Q moves with speed:**

$$a\sqrt{\left(\frac{\lambda}{3m}\right)}$$ **after impact.**

(c) The amplitudes of the motion of the particle P before and after impact are in the ratio $\sqrt{3}:1$.

(AEB)

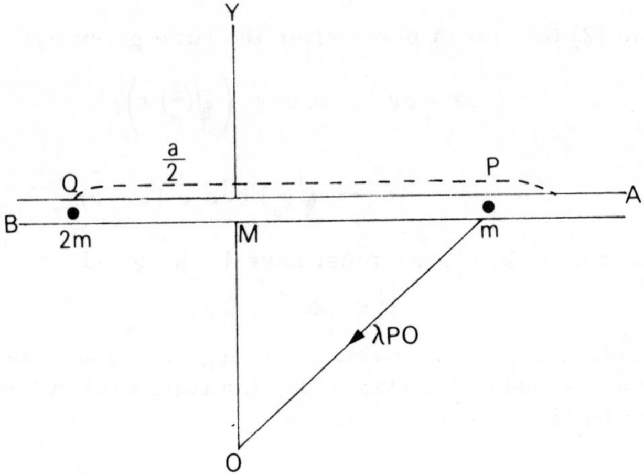

Choose M as origin, and denote the distance PM by x. Let OP = r. Then P is acted on by an attractive force λr towards O. This can be resolved into a component λr cos OPM = λx along PM, and one perpendicular to PM, which is balanced by the normal reaction. Hence the equation of motion of P in the direction MA is:

$$m \frac{d^2x}{dt^2} = -\lambda x$$

or:

$$\frac{d^2x}{dt^2} = \frac{-\lambda}{m} \cdot x = -\omega^2 x$$

where:

$$\omega^2 = \frac{\lambda}{m}$$

This is the equation of simple harmonic motion with M as centre. Since, also, P starts from rest when x = a, the amplitude is a. Therefore the velocity v of P when at any distance x from M is given by the usual equation:

$$v^2 = \omega^2(a^2 - x^2)$$

Hence if V be the velocity of P when it reaches Q, just before impact:

$$V^2 = \omega^2\left(a^2 - \left(\frac{a}{2}\right)^2\right) = \frac{3}{4}\omega^2 a^2 \qquad (1)$$

so: $$V = \frac{\sqrt{3}}{2}\,\omega a$$

Let V_1 and U be the velocities of P and Q respectively after impact both measured along MB. From the principle of conservation of momentum:

$$mV = mV_1 + 2m \cdot U$$

or: $$V = V_1 + 2U \qquad (2)$$

Since P and Q are perfectly elastic, e = 1, and so the law of impact gives:

$$U - V_1 = V \qquad (3)$$

From (2) and (3):

$$U = \frac{2V}{3} \qquad V_1 = \frac{-V}{3}$$

Therefore U, the velocity of Q after impact:

$$= \frac{2}{3} \cdot \frac{\sqrt{3}}{2}\,\omega a = a\sqrt{\left(\frac{\lambda}{3m}\right)}$$

on substituting for ω.

Let a_1 denote the amplitude of the motion of P after impact. The value of ω is unchanged so that:

$$V_1^2 = \omega^2\left(a_1^2 - (\frac{a}{2})^2\right)$$

i.e. $$\frac{V^2}{9} = \omega^2 (a_1^2 - \frac{a^2}{4})$$

Substitution for V^2 from (1) shows that, after reduction:

$$a_1^2 = \frac{a^2}{3}$$

i.e. $$a:a_1 = \sqrt{3}:1$$

36. **A vertical section of a valley is in the form of a parabola whose parametric equations with respect to horizontal and vertical axes OX, OY are $x = 2ap$, $y = ap^2$. A gun, situated at the point with co-ordinates $x = -2a$, $y = a$, fires a shell with velocity:**

$$\sqrt{(\frac{4ga}{3})}$$

at an angle of elevation α and in the plane OXY. Find the value of the parameter p at the point at which the shell strikes the ground. Deduce that, in order that the shell should hit a target situated at the lowest point of the valley, α must be chosen to be either:

$$\tfrac{1}{4}\pi \text{ or } \tan^{-1}\frac{1}{3}$$

(Ox)

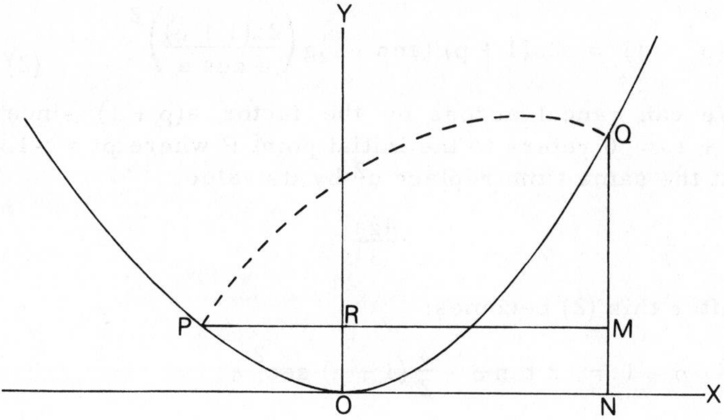

Let P be the point with co-ordinates $x = -2a$, $y = a$, i.e. the point for which $p = -1$. The origin O is the lowest point. Let the shell strike the ground at Q whose co-ordinates are $ON = 2ap$, $NQ = ap^2$. Draw PM parallel to OX meeting OY in R and QN in M. For the present denote:

$$\sqrt{\frac{4ga}{3}}$$

the initial velocity of the shell, by u. If t denotes the time the shell takes to reach Q, by the usual formulae for projectiles:

$$PM = ut\cos\alpha \quad ; \quad QM = ut\sin\alpha - \tfrac{1}{2}gt^2 \qquad (1)$$

Now: $PM = PR + RM = 2a + 2ap$

$$QM = QN - MN = ap^2 - a$$

Substituting in (1):

$$2a(1+p) = ut\cos\alpha \quad ; \quad a(p^2-1) = ut\sin\alpha - \tfrac{1}{2}gt^2$$

Eliminating t between these two equations:

$$a(p^2 - 1) = 2a(1 + p) \tan \alpha - \tfrac{1}{2}g\left(\frac{2a(1 + p)}{u \cos \alpha}\right)^2 \qquad (2)$$

We can cancel across by the factor $a(p + 1)$ since $p + 1 = 0$ refers to the initial point P where $p = -1$. At the same time replace u^2 by its value:

$$\frac{4ga}{3}$$

After this (2) becomes:

$$p - 1 = 2 \tan \alpha - \frac{3}{2}(1 + p) \sec^2 \alpha$$

from which:

$$p = \frac{2 + 4 \tan \alpha - 3 \sec^2 \alpha}{2 + 3 \sec^2 \alpha}$$

$$= \frac{4 \tan \alpha - 1 - 3 \tan^2 \alpha}{5 + 3 \tan^2 \alpha}$$

In order that the shell should hit a target situated at the lowest point O, the values given for p by this equation must be 0. Therefore:

$$4 \tan \alpha - 1 - 3 \tan^2 \alpha = 0$$

the roots of which for $\tan \alpha$ are 1 and 1/3. Hence α must be either:

$$\tfrac{1}{4}\pi \text{ or } \tan^{-1} \frac{1}{3}$$

37. Two small smooth inelastic spheres A and B, of equal radius and of masses m_1, m_2 respectively, rest at the edge of a smooth circular horizontal table, centre O and radius a, in such a way that $\angle AOB = 2/3\,\pi$. Initially both spheres are given the same velocity V towards O. Given that $m_1 \langle X_2$ and that the times which elapse between the impact and the moments at which A, B leave the table are t_1, t_2 respectively, prove that:

$$\frac{1}{t_1{}^2} - \frac{1}{t_2{}^2} = \frac{3V^2}{a^2}\left(\frac{m_2 - m_1}{m_2 + m_1}\right) e\ (1 + e)$$

where e is the coefficient of restitution between the spheres.

(Ox)

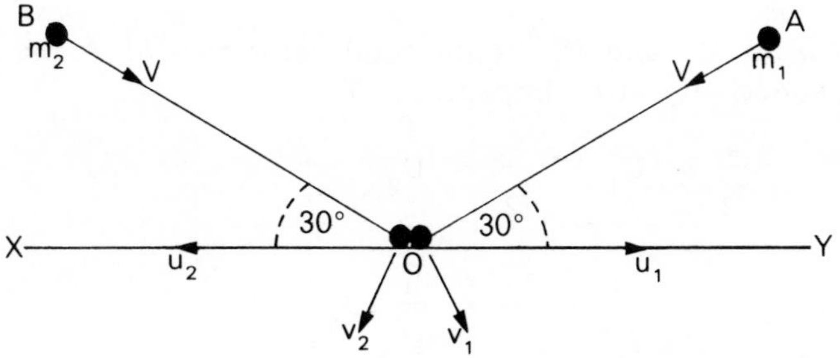

Let XOY bisect $\angle AOB$ externally, then $\angle XOB = \angle YOA = 30^o$. At the instant when the spheres collide at O, their centres will be along a line parallel to XOY. Let the velocities of A after impact be u_1 along OY and v_1 perpendicular to it. Let the velocities of B after impact be u_2 along OX and v_2 perpendicular to it.

Then from the law of conservation of momentum, in the direction OY:

$$m_1 u_1 - m_2 u_2 = (m_2 - m_1) V \cos 30^{o} \qquad (1)$$

From the law of impact:

$$u_1 + u_2 = 2eV \cos 30^{o} \qquad (2)$$

Also, since impact is smooth, velocities perpendicular to XOY are unaltered by impact, so that:

$$V \sin 30 = v_1 = v_2 \quad \text{(2 equations here)} \qquad (3)$$

Solving for u_1 and u_2 from (1) and (2):

$$u_1 = \frac{V \cos 30^{o} (2m_2 e + (m_2 - m_1))}{m_1 + m_2} \qquad (4)$$

$$u_2 = \frac{V \cos 30^{o} (2m_1 e - (m_2 - m_1))}{m_1 + m_2} \qquad (5)$$

Now let U_1 and U_2 be the total velocities of A and B respectively after impact. Then:

$$\frac{a}{t_1} = U_1 \; ; \; \frac{a}{t_2} = U_2$$

so:

$$\frac{1}{t_1^2} - \frac{1}{t_2^2} = \frac{1}{a^2}(U_1^2 - U_2^2) \qquad (6)$$

But:

$$U_1^2 = u_1^2 + v_1^2$$

and:

$$U_2^2 = u_2^2 + v_2^2$$

Consequently, since from (3) $v_1 = v_2$:

$$U_1^2 - U_2^2 = u_1^2 - u_2^2$$

Therefore from (6):

$$\frac{1}{t_1^2} - \frac{1}{t_2^2} = \frac{1}{a^2}(u_1^2 - u_2^2) = \frac{1}{a^2}(u_1 + u_2)(u_1 - u_2)$$

which from (4) and (5):

$$= \frac{3V^2}{4a^2} \cdot \frac{2(m_1 + m_2)e \cdot 2(m_2 - m_1)(1 + e)}{(m_1 + m_2)^2}$$

$$= \frac{3V^2}{a^2}\left(\frac{m_2 - m_1}{m_2 + m_1}\right) e\,(1 + e)$$

38. **A particle is projected at an angle of elevation a from a point O on a horizontal plane. The particle passes through a point P when travelling upwards at an angle θ to the horizontal . If the line PO is at an angle β to the horizontal, show that tan a, tan β, and tan θ are in arithmetic progression.**

The particle passes through points P_1 and P_2 for which $\beta = \beta_1$ and $\beta = \beta_2$ respectively. When passing through P_1 the particle is travelling upwards at an angle Ψ to the horizontal and when passing through P_2 is travelling downwards at an angle Ψ to the horizontal. If $\tan \Psi = \frac{3}{4}$, and $2 \tan \beta_2 = \tan \beta_1$, show than $\tan a = 2\frac{1}{4}$.

(AEB)

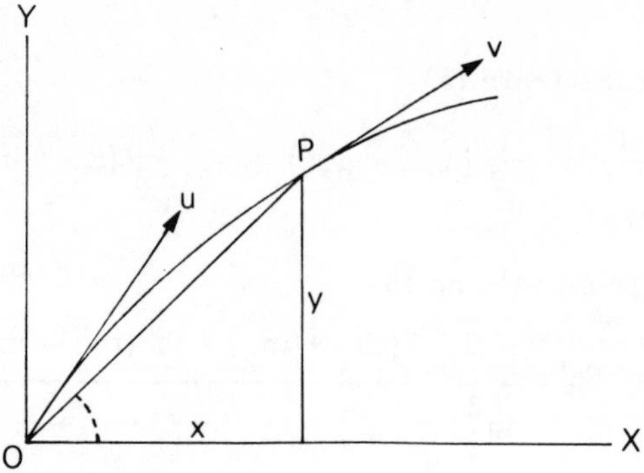

Let u be the initial velocity of projection of the particle from O, v its velocity when at P. Choose the horizontal and vertical lines through O in the plane of projection as x and y axes, and let the co-ordinates of P be x and y. Then, if t is the time the particle takes to reach P, from the usual formulae:

$$v \cos \theta = u \cos \alpha \qquad (1)$$

$$v \sin \theta = u \sin \alpha - gt \qquad (2)$$

$$x = u \cos \alpha . t \qquad (3)$$

$$y = u \sin \alpha . t - \tfrac{1}{2}gt^2 \qquad (4)$$

Dividing (2) by (1):

$$\tan \theta = \tan \alpha - \frac{gt}{u \cos \alpha}$$

Dividing (4) by (3):

$$\frac{y}{x} = \tan \beta = \tan \alpha - \frac{gt}{2u \cos \alpha} \qquad (5)$$

It follows that:

$$\tan\alpha - \tan\beta = \tan\beta - \tan\theta = \frac{gt}{2u\cos\alpha}$$

i.e. $\tan\alpha$, $\tan\beta$, $\tan\theta$ are in arithmetic progression. When the particle has reached P_1, $\beta = \beta_1$, and $\theta = \psi$. Hence $\tan\alpha$, $\tan\beta_1$, $\tan\psi$ are in arithmetic progression. Therefore, as $\tan\psi = \frac{3}{4}$:

$$\tan\beta_1 = \frac{\tan\alpha + \frac{3}{4}}{2}$$

But when the particle has reached P_2, after time t_2, it is travelling **downwards**, so that if v be its velocity now:

$$v\cos\psi = u\cos\alpha$$

but: $$v\sin\psi = gt2 - u\sin\alpha$$

Accordingly, from these:

$$\tan\psi = \frac{gt2}{u\cos\alpha} - \tan\alpha$$

Since: $$\tan\beta_2 = \tan\alpha - \frac{gt2}{2u\cos\alpha}$$

from (5), it follows that,

$$\tan\beta_2 = \frac{\tan\alpha - \tan\psi}{2}$$

$$= \frac{\tan\alpha - \frac{3}{4}}{2}$$

As $2\tan\beta_2 = \tan\beta_1$:

$$2\left(\frac{\tan\alpha - \frac{3}{4}}{2}\right) = \frac{\tan\alpha + \frac{3}{4}}{2}$$

$$\therefore \quad \tan\alpha = 2\tfrac{1}{4}$$

39. Two identical small smooth spheres A, B, are attached by equal light inextensible strings OA, OB, of length a, to a fixed point O on a smooth horizontal table. Initially, AOB is a straight line, the strings are taut, and A, B are moving with velocities u, 2u respectively in directions perpendicular to AOB such that the sense of rotation about O is the same for both A and B. Prove that, after 2n impacts, the velocities of A and B are:

$$\tfrac{1}{2}u(3 - e^{2n}) \quad ; \quad \tfrac{1}{2}u(3 + e^{2n})$$

respectively, and that the 2n th impact occurs after a time:

$$\frac{\pi a(2 - e^{2n-1} - e^{2n})}{u(1 - e)e^{2n-1}}$$

where $e(<1)$ is the coefficient of restitution between the spheres.

(Ox)

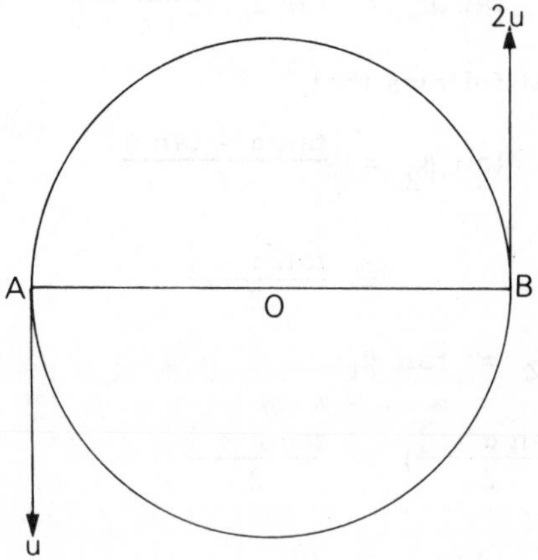

As the spheres and table are smooth, each sphere moves at constant speed along the circumference between impacts. Take the mass of each sphere as unity, and let u_n and v_n be the velocities of A and B respectively after the n th collision or impact between them. Then from the law of conservation of momentum:

$$u + 2u = u_1 + v_1 = u_2 + v_2 = \dots = u_n + v_n \qquad (1)$$

Also by Newton's law of impact:

velocity of separation = e x velocity of approach

Thus: $u_1 - v_1 = e(2u - u) = eu$, which implies that:

$$u_1 > v_1$$

$$v_2 - u_2 = e(u_1 - v_1) = e^2 u$$

so that: $v_2 > u_2$

Similarly:

$$u_3 - v_3 = e(v_2 - u_2) = e^3 u$$

$$v_4 - u_4 = e^4 u$$ and generally:

$$v_{2n} - u_{2n} = e^{2n} . u \qquad (2)$$

Also from (1):

$$u_{2n} + v_{2n} = 3u$$

Adding and subtracting these last two equations:

$$v_{2n} = \frac{u}{2}(3 + e^{2n})$$

$$u_{2n} = \frac{u}{2}(3 - e^{2n})$$

as was to be proved.

Before the first impact, B was overtaking A at a speed 2u - u = u, and, to do so, had to travel a distance πa. And so the first impact takes place after a time:

$$\frac{\pi a}{u}$$

Similarly the time between the first and second impacts is:

$$\frac{2\pi a}{u_1 - v_1} = \frac{2\pi a}{eu}$$

the time between the second and third impacts is:

$$\frac{2\pi a}{e^2 u}$$

and so on, the time between the (2_{n-1})th and 2n th impacts being:

$$\frac{2\pi a}{e^{2n-1} u}$$

Therefore, summing these, the 2n th impact occurs after a time:

$$\frac{\pi a}{u} + \frac{2\pi a}{eu} + \frac{2\pi a}{e^2 u} + \ldots + \frac{2\pi a}{e^{2n-1} u}$$

$$= \frac{\pi a}{u}(1 + \frac{2}{e} + \frac{2}{e^2} + \ldots + \frac{2}{e^{2n-1}} \quad (3)$$

Now the terms:

$$\frac{2}{e} + \frac{2}{e^2} + \ldots + \frac{2}{e^{2n-1}}$$

form a geometric series whose sum, by the formula, is:

$$\frac{2(1 - e^{2n-1})}{e^{2n-1}(1 - e)}$$

Hence, from (3), the 2n th impact occurs after a time:

$$\frac{\pi a}{u}\left(1 + \frac{2(1 - e^{2n-1})}{e^{2n-1}(1 - e)}\right) = \frac{\pi a(2 - e^{2n-1} - e^{2n})}{u(1 - e)e^{2n-1}}$$

40. **A smooth circular cylinder of radius a is fixed with its axis horizontal. Two particles, each of mass m, are connected by a light inelastic string of length $\pi a/3$ and rest in an equilibrium position on the surface of the cylinder in a vertical plane orthogonal to its axis. One of the particles is slightly displaced from this equilibrium position and the particles begin to slide on the surface of the cylinder in a plane perpendicular to its axis. Show that, when each particle has travelled a distance $a\theta$ on the surface of the cylinder, their velocity v is given by:**

$$v^2 = ga\sqrt{3}\,.\,(1 - \cos\theta)$$

and find the tension in the string at this point.

Show that one of the particles leaves the cylinder when each particle has travelled a distance $a\beta$ on the cylinder where:

$$\tan\beta = \frac{5\sqrt{3}}{11}$$

(Ox)

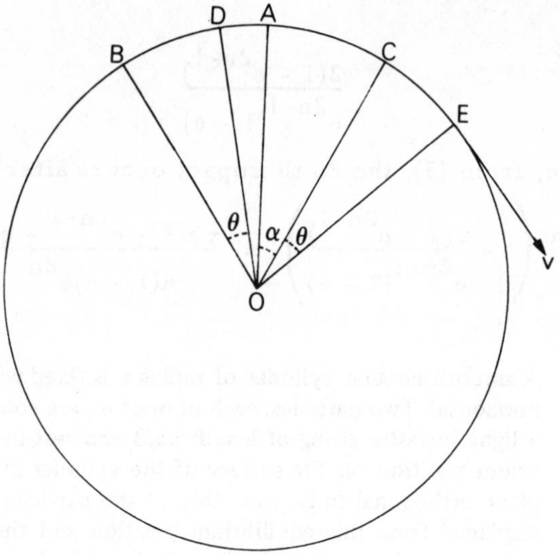

The diagram represents the vertical section of the cylinder in which motion takes place, O and A being the centre and highest point respectively of this section. Let B and C be the initial positions of the particles, D and E their respective positions after each has travelled a distance $a\theta$ on the cylinder, so that:

$$\angle BOA = \angle AOC = \frac{\pi}{6} = \alpha \text{ (for short)}$$

Also: $\angle BOD = \angle COE = \theta$

The particle which moves from C to E has fallen a vertical distance:

$$a(\cos\alpha - \cos(\alpha + \theta))$$

while the other, which moves from B to D, has risen a vertical distance:

$$a(\cos(\alpha - \theta) - \cos\alpha)$$

Hence by the principle of conservation of energy:

$$\tfrac{1}{2}mv^2 + \tfrac{1}{2}mv^2 = mga(\cos\alpha - \cos(\alpha+\theta)) -$$

$$mga(\cos(\alpha-\theta) - \cos\alpha)$$

so that:

$$v^2 = 2ga(\cos\alpha - \cos\alpha \,.\, \cos\theta)$$

on reducing:

$$= 2ga\cos\alpha\,(1 - \cos\theta)$$

$$= ga\sqrt{3}\,.\,(1 - \cos\theta) \qquad (1)$$

since: $\alpha = \dfrac{\pi}{6}$

Now let T be the tension in the string at this point. The velocity of the particle at E, namely v, $= a\dot{\theta}$ and its acceleration along the tangent to the circle is $a\ddot{\theta}$. From (1):

$$a^2 \,.\, \dot{\theta}^2 = ga\sqrt{3}\,.\,(1 - \cos\theta)$$

Differentiating with respect to t:

$$2a\dot{\theta}\,.\,\ddot{\theta} = \sqrt{3}\,.\,g\sin\theta\,.\,\dot{\theta}$$

or, on cancelling $\dot{\theta}$ which is not zero:

$$a\,.\,\ddot{\theta} = \frac{\sqrt{3}}{2}g\sin\theta \qquad (2)$$

The equation of motion for the particle at E along the tangent is thus:

$$ma\ddot{\theta} = mg\sin(\alpha+\theta) - T$$

so that, from (2):

$$T = mg(\sin(\alpha+\theta) - \frac{\sqrt{3}}{2}\sin\theta)$$

$$= \tfrac{1}{2}mg\cos\theta \text{ on reducing}$$

since: $\sin \alpha = \frac{1}{2}$; $\cos \alpha = \frac{\sqrt{3}}{2}$

To find when one particle leaves the cylinder let R be the pressure of the cylinder on the particle at E. This, as the cylinder is smooth, will be along OE. The acceleration of this particle along EO is:

$$\frac{v^2}{a}$$

Forming its equation of motion along EO:

$$m\frac{v^2}{a} = mg \cos(\theta + \alpha) - R$$

so: $$R = mg \,.\, \cos(\theta + \alpha) - m\frac{v^2}{a}$$

$$= mg(\cos(\theta + \alpha) - \sqrt{3}(1 - \cos\theta))$$

from (1).

From this it follows that R steadily decreases as θ increases but, being a pressure, can never be negative. Consequently when R = 0, this particle leaves the cylinder, that is, when:

$$\cos(\theta + \alpha) - \sqrt{3}(1 - \cos\theta) = 0$$

or: $$\frac{3\sqrt{3}}{2}\cos\theta - \frac{\sin\theta}{2} - \sqrt{3} = 0$$

$$3\sqrt{3}\cos\theta - \sin\theta = 2\sqrt{3}$$

To find tan θ, divide across by cos θ and square both sides:

$$(3\sqrt{3} - \tan\theta)^2 = 12(1 + \tan^2\theta)$$

$$\therefore \quad 11 \tan^2 \theta + 6\sqrt{3} \tan \theta - 15 = 0$$

$$(11 \tan \theta - 5\sqrt{3})(\tan \theta + \sqrt{3}) = 0$$

This gives: $\tan \theta = \dfrac{5\sqrt{3}}{11}$

the second value $\tan \theta = -\sqrt{3}$ not being admissible. With this value for θ, one particle leaves the cylinder after each particle has travelled a distance $a\theta$ on it.

41. **A particle of mass m has the velocity vector 3i + 2j when at the point P whose position vector is i + j; two seconds later it has the velocity vector 7i — 10j and is at Q. The particle is constrained to move along the curve:**

$$\mathbf{r} = (a_1t^2 + a_2t + 1)\mathbf{i} + (b_1t^2 + b_2t + 1)\mathbf{j}$$

where t seconds is the time since the particle was at P. Find the values of the constants a_1, a_2, b_1 and b_2 and show that the force acting on the particle is constant.

Calculate, giving your answer as a vector, the moment of the force about the origin when the particle is at P, and find the work done by the force in moving the particle from P to Q.

(AEB)

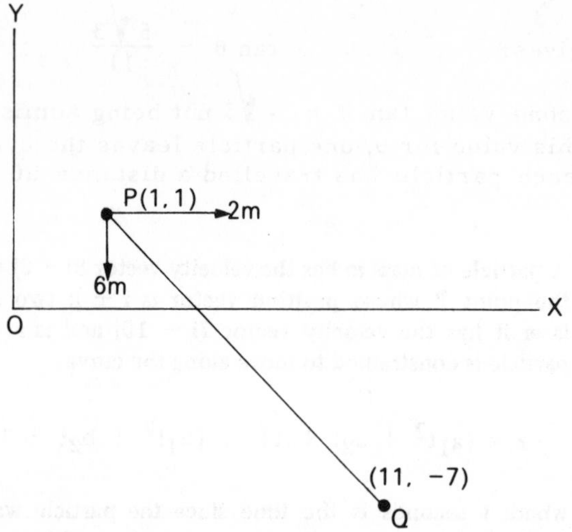

Note to reader. This diagram anticipates some of the results proved in what follows.

Differentiating r with respect to t, the velocity vector is seen to be:

$$(2a_1t + a_2)i + (2b_1t + b_2)j \qquad (1)$$

When t = 0, this becomes $a_2i + b_2j$, the velocity vector at P. Thus $a_2 = 3$, $b_2 = 2$. When t = 2, (1) becomes:

$$(4a_1 + a_2)i + (4b_1 + b_2)j$$

the velocity vector at Q. Hence $4a_1 + a_2 = 7$; $4b_1 + b_2 = -10$, so that: $a_1 = 1$; $b_1 = -3$. Differentiating (1), the acceleration vector is:

$2a_1 . i + 2b_1 . j$; i.e. $2i - 6j$, which is constant.

Since force = mass x acceleration, the force acting on the particle is constant, with components 2m and -6m parallel to the x and y axes.

When the particle is at P the moment of this force about the origin O is 2m x 1 + 6m x 1 = 8m in a clockwise, or negative, sense about O. Since the axis of rotation is perpendicular to the xy plane, it is the z axis. Therefore the moment of this force is, in vector form, -8m . k.

Now letting t = 2, Q is the point:

$$(1 \cdot 2^2 + 3 \cdot 2 + 1)i + (-3 \cdot 2^2 + 2 \cdot 2 + 1)j \qquad \text{or:}$$

$$11i - 7j$$

Therefore the work done by the force in moving the particle from P to Q is:

$$2m(11 - 1) + 6m(7 + 1) = 68m$$

42. **A uniform rod AB of mass M and length 2a is freely pivoted at its mid-point. When the rod is at rest in a horizontal position, it is struck at B by a particle of mass m falling vertically with speed u. The coefficient of restitution between the particle and the rod is e and the particle is reduced instantaneously to rest by the impact. Prove that:**

$$\frac{m}{M} = \frac{e}{3}$$

and find, in terms of a, e, and u, the time taken for the rod to perform a complete revolution after impact.

(Ox)

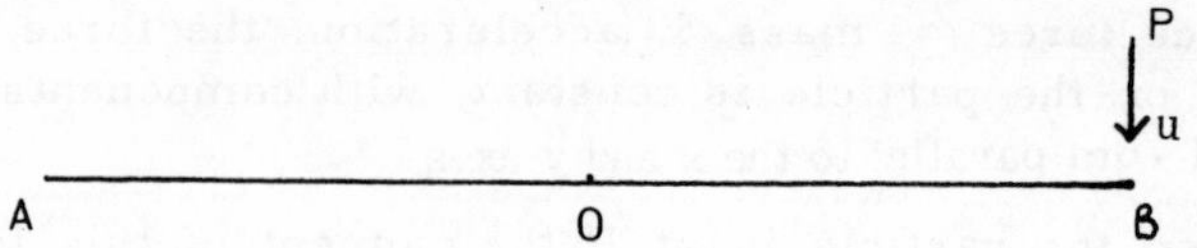

Let O be the mid-point of the rod, P the particle. At the instant of impact P has momentum mu, and so mu is the impulse received by the rod. Let ω be the angular velocity of the rod after impact. The principle used now is that the change in the moment of momentum of the rod about any fixed axis L is equal to the moment of the impulse received by the rod about L. Take L to be the horizontal line through O perpendicular to AB. Then since the moment of inertia of the rod about L is:

$$M \frac{a^2}{3}$$

and the moment of the impulse about L is mu . a:

$$M \frac{a^2}{3} (\omega - 0) = mu . a$$

or: $$Ma\,\omega = 3mu \qquad (1)$$

Also, by the law of impact:

velocity of separation = e x velocity of approach

$$a\,\omega - 0 = e(u - 0)$$

or: $$a\,\omega = eu \qquad (2)$$

Division of (1) by (2) gives:

$$M = \frac{3m}{e}$$

i.e. $$\frac{m}{M} = \frac{e}{3}$$

Since the rod is freely pivoted at O, only forces which pass through O, and therefore meet L, act on the rod after impact. Its rate of change of moment of momentum about L is, therefore, zero, i.e. its angular velocity is constant, equal to ω. The time taken to perform a complete revolution is thus:

$$\frac{2\pi}{\omega} = 2\pi\frac{a}{eu}$$

from (2).

43. A car of mass m is moving in a straight line on a rough horizontal plane. At time t the car is moving with velocity v, and the resistance to its motion is kv where k is a constant. If the car works at a constant rate h, show that:

$$mv\frac{dv}{dt} + kv^2 = h$$

If the car starts from rest, show that v is always less than:

$$\left(\frac{h}{k}\right)^{\frac{1}{2}}$$

and find the time taken for the car to reach the speed:

$$\frac{1}{2}\left(\frac{h}{k}\right)^{\frac{1}{2}}$$

(Ox)

If P is the force exerted by the engine at time t, then:

$$Pv = h \qquad (1)$$

Also, by Newton's second law of motion, if f is the acceleration at the instant, then:

$$P - \text{resistance} = mf = m\frac{dv}{dt}$$

$$\therefore \qquad P - kv = m\frac{dv}{dt}$$

so:

$$P = kv + m\frac{dv}{dt}$$

Accordingly, substituting this for P in (1):

$$mv\frac{dv}{dt} + kv^2 = h$$

or:

$$\frac{mv\frac{dv}{dt}}{h - kv^2} = 1$$

Integrate with respect to t:

$$-\frac{m}{2k}\log_e(h - kv^2) = t + C$$

where C is a constant to be found.

When t = 0, v = 0, and so:

$$C = -\frac{m}{2k}\log h$$

$$\therefore \qquad t = \frac{m}{2k}(\log h - \log(h - kv^2))$$

$$= \frac{m}{2k}\log\left(\frac{1}{1 - \frac{k}{h}v^2}\right)$$

Since $t > 0$, $1 - \frac{k}{h}v^2$ must be positive, so that:

$$v < \left(\frac{h}{k}\right)^{\frac{1}{2}} \text{ always.}$$

When:

$$v = \frac{1}{2}\left(\frac{h}{k}\right)^{\frac{1}{2}}$$

$$t = \frac{m}{2k} \log \left(\frac{1}{1 - \frac{k}{h} \cdot \frac{h}{4k}}\right)$$

$$= \frac{m}{2k} \log_e \left(\frac{4}{3}\right)$$

44. Two stations A and B are 5·1 km apart. A train starts from A, accelerates uniformly till it reaches a speed of 20 metres/s, and travels at this speed for some time. Then, the brakes being applied, it decelerates uniformly till it stops at B. If the total time taken from A to B is 5 minutes, and the train takes 10 seconds longer to decelerate than to accelerate, find:

(a) the time taken to accelerate.

(b) the time during which the train travels at constant speed.

(c) the acceleration and deceleration (retardation).

Let f metres/s^2 and f_1 metres/s^2 be the acceleration and deceleration respectively. Let t seconds be the time of acceleration, so that (t + 10) seconds will be the time of deceleration. Thus from the given conditions (total time taken is 300 seconds):

$$(300 - 2t - 10) \text{ seconds} \qquad (1)$$

will be the time during which the train travels at constant speed.

Using the formula:

$$v = u + f\,t$$

we get: $f\,t = 20 = f_1(t + 10)$ (2)

The distance travelled during acceleration $= \frac{1}{2}f \cdot t^2$

$$= \tfrac{1}{2}t \times f\,t = \tfrac{1}{2}t \times 20 \text{ m}$$

Similarly the distance travelled during deceleration:

$$= \tfrac{1}{2}(t + 10)\ 20 \text{ m}$$

Also distance travelled at constant speed is, from (1):

$$20(300 - 2t - 10) \text{ m}$$

Hence, adding:

$$20\frac{t}{2} + 20\frac{(t + 10)}{2} + 20(300 - 2t - 10) = 5100 \text{ m}$$

or simplifying:

$$t + 5 + 300 - 2t - 10 = 255$$

which gives:

$$t = 40 \text{ seconds}$$

Substitution for t in (2) gives:

$$f = \tfrac{1}{2} \text{ metre/s}^2$$

$$f_1 = \frac{2}{5} \text{ metre/s}^2$$

Answers: (a) 40 seconds

(b) 210 seconds

(c) $\frac{1}{2}$ metre/s^2 and $\frac{2}{5}$ metre/s^2

45. **Two light inelastic strings AP and BP connect a particle P to fixed points A and B. The point B is vertically above A and AB = AP = ℓ, and BP = ℓ . √3. The particle P moves in a horizontal circle with constant speed. The least angular speed of P for both strings to be taut is ω. At this speed calculate the angle between the strings and the value of ω.**

When the angular speed of P is ω_1 (> ω) the tensions in the strings are equal. Show that:

$$\omega_1^{\ 2} = \frac{2g}{(1\sqrt{3})}$$

(AEB)

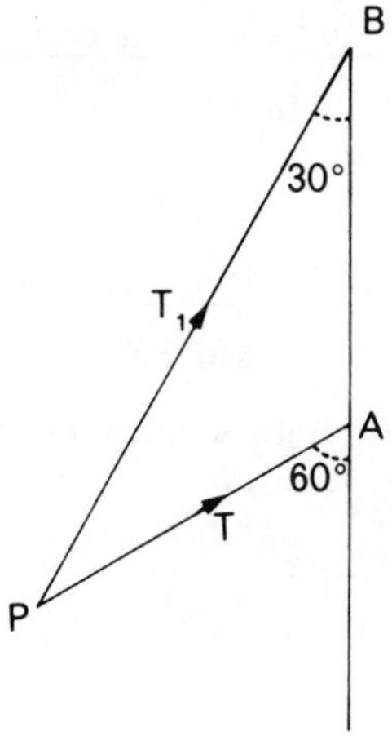

When both strings are taut, since AB = AP = l, and BP = l √3, it can be shown, from the cosine formula for ∠PAB, or otherwise, that ∠PAB = 120°, so that ∠APB = ∠ABP = 30°.

The particle moves in a horizontal circle whose radius will be l cos 30, with angular speed ω, and hence with an acceleration l cos 30° . ω^2 towards the centre of the circle.

Let T and T_1 be the tensions in PA and PB respectively. Resolving for the equilibrium of P, vertically:

$$T \sin 30^\circ + T_1 \sin 60^\circ = mg \qquad (1)$$

Resolving horizontally, the equation of motion of P is:

$$T \cos 30^\circ + T_1 \cos 60^\circ = ml \cos 30^\circ . \omega^2 \qquad (2)$$

Solving (1) and (2) for T and T_1:

$$T = \frac{m(l \cos 30^\circ \sin 60^\circ . \omega^2 - g \cos 60^\circ)}{\sin 60^\circ \cos 30^\circ - \cos 60^\circ \sin 30^\circ}$$

$$= \frac{m(l \cos^2 30^\circ . \omega^2 - g \cos 60^\circ)}{\sin 30^\circ}$$

Similarly:

$$T_1 = \frac{m(g \cos 30^\circ - l \cos 30^\circ \sin 30^\circ . \omega^2)}{\sin 30^\circ}$$

In order that both strings should be taut, T and T_1 must not be negative, and so:

$$\omega^2 \geqslant \frac{g}{l} . \frac{\cos 60^\circ}{\cos 30^\circ \sin 60^\circ}$$

i.e. $\omega^2 \geqslant \frac{2}{3} . \frac{g}{l}$

Also:

$$g \geqslant l \sin 30^\circ . \omega^2$$

so: $\omega^2 \leqslant \frac{2g}{l}$

From the first of these, the least angular speed for both strings to be taut is:

$$\sqrt{\frac{2g}{31}}$$

When the angular speed of P is ω_1 ($> \omega$) the tensions in the strings are equal. This means that:

$$1 \cos^2 30^\circ \quad \omega_1^2 - g \cos 60^\circ$$

$$= g \cos 30^\circ - \ell \cos 30^\circ \sin 30^\circ \,.\, \omega_1^2$$

$$\omega_1^2 = \frac{g(\cos 60^\circ + \cos 30^\circ)}{1 \cos 30 (\cos 30 + \sin 30)} = \frac{2g}{1\sqrt{3}}$$

NOTES

I hate Applied Maths, & if I don't pass the exam, I don't give a damn.

NOTES

NOTES

Multiple Choice A Level Objective tests are now an important part of most A Level examinations. This series presents batteries of common questions and is also an excellent way of revising essential facts. Subjects covered in this series are: Pure Mathematics, Applied Mathematics, Chemistry, Physics, and Biology.

Test Yourself A series of pocket books designed for the revision of essential facts whenever the student has a free moment. Subjects covered in this series are: English, Language, French, German, Commerce, Economics, Mathematics, Modern Mathematics, Chemistry, Physics, Biology, and Human Biology.

Celtic Revision Aids can make the difference between passing or failing your examination.